D0977654

Praise for *Queen Bess:*

"What makes *Queen Bess: Daredevil Aviator* intriguing is how this Chicago manicurist, born dirt-poor in Atlanta, Texas, in 1892, became America's first black aviatrix. . . . Bessie Coleman crammed a lot of living into her thirty-four years. Her spirit of adventure, coupled with dogged determination to break down racial barriers in the United States, has inspired other African American women and men to dare to dream."
> —*Quarterly Black Review of Books*

"The definitive study of a remarkable young African American woman."
> —*Choice*

"A fascinating story."
> —*Antiques and Americana*

"Doris L. Rich . . . brings Bess Coleman to life and bestows her legacy with the recognition she rightfully deserves."
> —*Aviation*

"I point to Bessie Coleman and say without hesitation that here is a woman who exemplifies and serves as a model to all humanity: the very definition of strength, dignity, courage, integrity, and beauty."
> —Mae Jemison

"A timely and engaging introduction to a woman of stunning accomplishment and courage who deserves a place of high honor in the pantheon of early flying."
> —*Kirkus Reviews*

"A gem."
> —*Aerospace*

"A vivid portrait of a woman who bucked tradition, racial prejudice, and sexual discrimination in her determination to take to the skies."
> —*Chicago Tribune*

"*Queen Bess* is a valuable contribution to four kinds of history—of blacks, of women, of Chicago, and of aviation. . . . Bessie Coleman was a great trailblazer, single-handedly opening the frontier of aviation for blacks. The Tuskegee Airmen, the famous black fighter pilots of World War II, can trace their beginnings to her example."
> —Henry Kisor, *Chicago Sun-Times*

Queen Bess

Daredevil Aviator

Doris L. Rich

Smithsonian Institution Press

Washington and London

This book was edited by Initial Cap Editorial Services and designed by
Janice Wheeler.

The paper used in this publication meets the minimum requirements of the
American National Standard for Permanence of Paper for Printed Library Materials
Z39.48-1984.

Library of Congress Cataloging-in-Publication Data

Rich, Doris L.
    Queen Bess : Daredevil aviator / Doris L. Rich.
        p.   cm.
Includes bibliographical references and index.
ISBN 1-56098-265-9 (alk. paper)
1. Coleman, Bessie, 1896–1926.   2. Afro-American air pilots-
-Biography.   3. Women air pilots—United States—Biography.
I.   Title.
TL540.CB46R52   1993
629.13′092–dc20
[B]   93-14785
                                                                      CIP

Printed in the United States of America.

10 9 8 7 6 5 4   06 05 04 03 02 01

For permission to reproduce individual illustrations appearing in this book, please
correspond directly with the owners of the images, as stated in the picture captions.
The Smithsonian Institution Press does not retain reproduction rights for these
illustrations individually or maintain a file of addresses for photo sources.

FOR CELIA AND RUBÉN

Because of Bessie Coleman, we have overcome that which was much worse than racial barriers. We have overcome the barriers within ourselves and dared to dream.

Lt. William J. Powell
Founder, Bessie Coleman Aero Clubs

Note numbers are not used in this book. Instead, as a pleasurable convenience to the reader, notes are printed at the back of the book and are identified by page number and an identifying phrase or quotation from the text.

CONTENTS

ACKNOWLEDGMENTS

Bessie Coleman in life was a challenge to the status quo—a constant struggle against the myriad of limitations conventional society erected against anyone who dared to be different. In death she presents an equal challenge to the biographer, for she left little of the material necessary for a reconstruction of her life.

African American intellectuals of her day left behind published works, minutes of meetings, correspondence, reminder notes on scratch pads and calendars for would-be literary detectives to examine. Similarly, black society of her time—many educated at Howard, Spelman, or Hampton—handed down a wealth of miscellaneous source material such as wills, financial records, photograph albums, and invitations to weddings, debuts, benefit balls, dinner parties, and graduation ceremonies. A Chicago manicurist raised in the cotton fields of Texas, Bessie read the works of the former and did the nails of the latter but belonged to neither group. She left behind virtually no personal memorabilia and all but a few of the letters she wrote have been lost. Thus the biographer attempting to reconstruct her life is primarily dependent on two sources—the recollections of her contemporaries and reports in the media.

Each source holds both rewards and limitations.. The brief unpublished memoirs of her oldest sister Elois provided an invaluable starting point. But Bessie was born a century ago and the few people who knew her and are still alive today are now in their seventies and older. While many of their remembrances are vivid, precise dates and places are often vague and conflicting.

"The media" in Bessie's case essentially means black weekly newspapers. The white press generally ignored Bessie, as they did most blacks unless they were actors or athletes, or involved in sex, crime, or violence. Fortunately, however, Bessie was considered an entertainer (aviation was yet to be elevated to a profession) and a celebrity. So the black media interviewed her and covered her activities. Microfilm and a few originals of these newspapers were a principal source of information on Bessie's career.

Other material in this book derives from special collections and the archives of the following libraries and special collections: Library of Congress, Martin Luther King Memorial Library, and the Moorland-Spingarn Research Center at Howard University, all in Washington, D.C.; Black Archives Research Center and Museum, Florida A&M University; Chicago Historical Society; City of Dallas Public Library; Du Sable Museum of African-American History, Chicago; Eartha White Collection, Thomas G. Carpenter Library, University of North Florida; Ellis County Museum, Inc.; Eugene C. Barker Texas History Center, University of Texas at Austin; Florida Collection, Jacksonville Public Libraries; Fullen Library, Georgia State University; Houston Metropolitan Research Center, Houston Public Library; Jacksonville Historical Society; Lilly Library, Indiana University; Memphis Shelby County Public Library; Microfilms Collection, Pennsylvania State University; Nicholas Sims Library in Waxahachie; Rosenberg Library, Galveston Public Library; Schomburg Center for Research in Black Culture, New York Public Library; Soper Library, Morgan State University; State Historical Society of Missouri; State Historical Society of Wisconsin; William Henry Smith Memorial Library, Indiana Historical Society. In all of them the staffs were patient and helpful. I wish to thank in particular Eileen D. Brady, Brenda T. Brown, Theresa Christopher, Barbara W. Clark, Jan Curry, Brian Dirck, Joan Dobson, James N. Eaton, Ralph L. Elder, Wilma Gibbs, Maureen Hady, C. H. Harris, Ara Kaye, Reba King, Margaret Koyne, Gary F. Kurutz, Patricia M. LaPointe, Genette McLaurin, Shirlene Newman, Shannon Simpson, Janet Sims-Wood, Martha Smith, Stephen C. Stappenbeck, Saundra Taylor, and Kenneth J. Whisenton.

During the two years that I worked on this book I spent many days in the library of the National Air and Space Museum of the Smithsonian Institution where I was given advice and encouragement by Tim Cronen, Tom Crouch, Robert Dreeson, Phil Edwards, Dan Hagedorn, Melissa Keiser, Mary Pavlovich, David Spender, Patricia Williams, and Larry Wilson. I thank them all.

I am especially grateful for the interviews given me by Bessie Coleman's nieces, Marion Coleman and Vera Buntin; her nephews, Dean Stallworth and Arthur Freeman; and other members of the family, Vera Jean Ramey, Jilda Motley, Fredia Delacoeur, and Alberta Lipscombe.

Others who gave me interviews were Inez Bentley, Mattie Borders, Eugenia Brown, A'Lelia Bundles, Dr. Margaret T. Burroughs, Cornelius R. Coffey, Patricia Fletcher, Beulah Florence, Jean Albright Gilley, George E. Haddaway, Bernice Hamilton, B. K. Hargrove, Ben J. Henderson, John P. Ingle, Marion Jeffers, Frances Johnson, William T. Johnson, Doris Jones, Dr. Marjorie Stewart Joyner, Georgia Lewis, Neal Loving, John C. McDonald, Hettie Mills, Mrs. W. C. Pittman, Annie Pruitt, G. Edward Rice, Theodore W. Robinson, Jacqueline Smith, Arthur Spaulding, Audrey Tillinghast, Camilla Thompson, Rutha Waters, Seth Williams, and Jean Yothers.

Information through correspondence came from Dr. Johnnetta

Betsch Cole, Georgia Conlon, Dr. Gabriele Dankert, Frances E. Davenport, Marie-Josèphe de Beauregard, Diana Estudillo, Stuart L. Faber, Doris H. Farr, H. Oakley Haynie, Joan Hrubec, Lenore Kieiling, Barbara Kozuh, Edward L. Leiser, Paul McCully, Wolfram Müller, Dr. Lorna Marie Polk, Deborah Palacios, Henry Snyder, Bernice T. Steadman, Victoria K. Steele, Dempsey J. Travis, and John Underwood.

Paulette Floyd, C.S.W., read early versions of the manuscript and provided valuable suggestions.

Offering shelter and good advice during research trips were Maria and Malcolm Bellairs, Carol Covington, Barbara Gault Hayes, Mary and William North, Claudia Oakes, Theresa and N. G. "Pat" Payne, Chris Rich, and D. Vicky Spencer-Burrows.

I am indebted to Dr. Mari Evans for the conclusion of my story. Only a poet as gifted as she could evoke the essence of the irrepressible Bessie.

For their editorial guidance I thank Felix C. Lowe, Ruth Spiegel, and Therese D. Boyd. And finally my thanks to Stanley Rich, my husband, who ferreted out flawed writing with gimlet-eyed efficiency but spoke of it gently to this writer. He should have been named co-author but modestly declined.

Queen Bess

CHAPTER 1

The Reluctant
Cotton Picker

On October 15, 1922, eight-year-old Arthur Freeman stood at the edge of a runway at Chicago's Checkerboard Airdrome, his head thrown back, looking wide-eyed into the sky at the Curtiss Jenny performing a figure eight. At the top of the eight the plane seemed suddenly to heel over and plunge downward, gaining speed as it hurtled toward earth. Just 200 feet above the runway, the aircraft slowed, shuddered, then slowly nosed up, soaring back into the sky before circling the field and coming in for a perfect landing.

"That's my aunt!" Arthur shouted.

It's doubtful anyone heard him. The 2,000 people in the bleachers right behind him were making enough noise of their own, a din of yelling, clapping, and whistling as the plane rolled to a stop and the pilot climbed down from the cockpit, pushed oil-smeared goggles up over a leather helmet, and smiled at them.

Arthur's aunt was Bessie Coleman. the first African American to earn an international pilot's license and the first black woman in the world to fly an airplane.

For a moment she stood by the plane, her long coat open, her uniform tailored like those worn by Canadian aviators of World War I. Puttees, cloth strips wound spirally around her legs from ankle to knee, topped her shiny leather boots, and a Sam Browne officer's belt, with its over-the-shoulder strap, circled her trim waist. Not only did Arthur's aunt know how to fly. She was also beautiful and small with a perfect figure and a flair for dramatic dress.

Bessie beamed at the adoring crowd, savoring the fact she had finally made it home in triumph as she had always said she would.

It had not been easy. Born in a one-room cabin, raised in a single-parent family, and educated in a school for black children that closed whenever the cotton needed picking, she went from doing laundry in Waxahachie, Texas, to manicuring nails in Chicago. Her decision to learn to fly presented more obstacles than those she had already faced. Black aviators couldn't teach her because there weren't any and white aviators wouldn't. But, someone told her, if her own countrymen couldn't help her others would. After learning French at a Michigan Avenue language school, Bessie made her way to Europe and earned a flying license from the Fédération Aéronautique Internationale, honored and recognized by every nation in the world.

Now, finally, she was home. "I've shown them all," she thought, supporters and critics, believers and doubters alike. She had proved herself to all Chicago, the South Side of her own people and the white world outside. As she walked proudly toward the stands she winked at Arthur and the rest of her family who had come to see her perform. The black weekly Chicago Defender *had already proclaimed her "Queen Bess." She had proved herself to all Chicago. Now she would prove herself to the rest of the world.*

The baby was a girl, tiny but perfectly formed, a beautiful copper color. Her little legs and arms waved vigorously, and her cries filled the air of the one-room, dirt-floored cabin as her mother wrapped her in a worn cotton quilt and laid her on the straw-filled mattress. With a sigh of relief Susan Coleman looked down at the infant. A healthy one, she thought. I can get back to work by morning.

The tall, slim black woman whose handsome face was marked more by care than laughter was no stranger to childbirth. She married in 1875 at age 20, later than most black women in the South. In the following twenty-three years she had thirteen children, four of whom did not survive childhood.

The baby Susan had just laid on the bed, born on January 26, 1892, was named Bessie. Her arrival was not recorded on a birth certificate or in a family Bible. Neither Susan nor her husband, George Coleman, could read or write. It is not known if Bessie's parents, both born in Texas, were slaves before the Civil War. It is likely that Susan was because both of her parents were originally from Georgia, migrating to Texas before her birth. George Coleman, from whom Bessie inherited her copper-colored skin, may have been born free. Three of his grandparents were Indian— probably Choctaw or Cherokee. George's father was from Indian territory, his mother from Missouri.

When Bessie was born, George and Susan were living in Atlanta, Texas, less than ten miles west of where the borders of Texas, Arkansas, and Louisiana meet. The town had been named by migrating Georgians for the city they had left when they traveled west to build the Texas and Pacific Railroad. Although it was founded only three years before the Colemans were married in 1875 and its residents numbered fewer than 1,000, Atlanta was a place where fortunes could be made in railroads, oil, and lumber.

On the main street, shaded by pin oaks arching overhead, citizens gathered at the general store run by R. S. Allday, who had opened his "mercantile establishment" before the railroad had reached the city, back when goods were hauled by ox cart from the river boats that docked at Jefferson, twenty miles southeast of Atlanta. The town's shady main street and general store were not part of the Colemans' world, however. Theirs was one of dirt roads at the edge of town, of tenant farms, and tiny, crowded cabins and incessant labor.

This was the world Bessie entered, one not only of poverty but of repressed rage and fear. Two years before her birth the state of Mississippi had begun the process of disenfranchising African Americans by legal means, a process soon followed by all of the Southern states. Three months after Bessie's birth a black postal employee in Memphis, Tennessee, and his two partners who had financed a small grocery store were taken by a mob from a Memphis jail and killed a mile outside the city. Their offense was to defend their property from an armed attack by white night raiders.

A year later, at a lynching in Paris, Texas, a black man accused of raping a five-year-old girl was first tortured with red-hot irons, then burned alive. Schoolchildren were excused from their classrooms to witness the burning and the railroads ran excursions for spectators from nearby communities. After the body was reduced to ashes, a mob fought over the bones, teeth, and buttons. Before Bessie was a year old, this and similar incidents had launched crusading black journalist Ida B. Wells into an investigation of lynchings throughout the South—728 of them in a single decade.

African Americans were not the only victims of violence and injustice. In Texas, for example, violence was common in both the black and white communities. In the last week of January 1892, Dallas—then a city of 40,000—was the scene of eight murders, three fires, two rapes, and four robberies. At the same time, the entire nation was plagued by a severe economic depression. High tariffs enabled millionaire industrialists to create monopolies in trade and industry while laborers worked twelve-hour days, seven days a week. When the workers unionized and called strikes, the industrialists hired armed men as strike breakers. In March of 1893, when Bessie was fourteen months old, panic in the financial centers of the country sent stocks plunging. Banks failed. Farmers lost their land. Prices fell, factories closed, and unemployment increased.

Somehow, in spite of the nationwide depression, hardworking, frugal George Coleman managed to save enough money from his wages as a day laborer to buy a small plot of land in Waxahachie, thirty miles south of Dallas. George Coleman may have chosen Waxahachie, a town of fewer than 4,000 inhabitants, because he saw an opportunity there to make more money than he had been making in Atlanta. The seat of Ellis County, self-proclaimed "largest cotton-producing county in the United States," Waxahachie somehow prospered and remained relatively untouched by the national depression. Oil had not yet come to dominate the Texas economy; cotton was still king. Straddled by two railroads—the Fort Worth and New Orleans, and the Missouri-Kansas-Texas (which would become the Chicago–Rock Island)—Waxahachie was a teeming hub of cotton yards, cotton warehouses, and cotton mills.

As in most cities and towns of the South, housing was segregated, the railroad tracks dividing the west side, where the whites lived, from the east, where the property George Coleman bought was located. Ellis County records show that on October 15, 1894, Coleman bought "one quarter of an acre, more or less," on Mustang Creek. A justice of the peace and notary public, Y. D. Kemble, notarized the bill of sale on January 1, 1895, "at 2½ o'clock P.M." The price was twenty-five dollars. The lot was four miles from the center of town on Palmer Road, near the railroad tracks on the east side where black workers had settled along East Main and Wyatt streets. Here small homes, single-cell and two- and three-room houses, lined the unpaved streets. It was more or less a separate community within Waxahachie where blacks established their own religious, commercial, and social institutions.

On his quarter acre, a wedge of land fronting sunbaked Palmer Road before plunging in the rear to tree-shaded Mustang Creek, Bessie's father built a three-room "shotgun" dwelling, characterized by a series of doors from one room to the next through which one could "shoot a shotgun the length of the house." It was Tshaped, with the rectangular front room placed crosswise, the back two lengthwise, and a porch added to both front and back.

Bessie was two when the family moved into the new dwelling. Her early childhood was a happy one, spent playing on the front lawn edged by red and yellow roses her mother had planted. When it rained she took cover on the front porch and watched small puddles form near the steps below. These would be her "lakes" for sailing leaf boats and later there would be mud for making "marbles" and mud pies.

Sundays were spent at church, morning and afternoon. Awakened at dawn, bathed and dressed in her best frock, Bessie was handed over to one of her two older brothers, Isaiah or John, to be kept clean and out of mischief until Susan was ready. The church was African American Baptist. If anything like the one Susan chose years later in Chicago, it was conservative. Members might sing enthusiastically but there would be no dancing in the aisles. To Susan Sundays were the Lord's and her children were to worship the Lord as she did, frequently and with great respect.

In 1894, soon after the Colemans moved to Waxahachie, Susan had another daughter, Elois. Two more girls followed, Nilus, born in 1896, and Georgia, in 1898. As John and Isaiah began to find work in the fields, Bessie was expected to keep an eye on her sisters, a responsibility assigned the eldest girl in many rural families. She also watered her mother's plants on the front porch and weeded the back garden where Susan raised corn, peanuts, and vegetables. The pear, peach, and plum trees were too tall for her to climb but she harvested the windfall fruit as it ripened.

There were also household chores for her in a family that numbered eight by 1900. Three of the older children no longer lived at home. Lillah, age 25, Alberta, 21, and Walter (also known as "Bud" or "Samy"), 17, had all left the family home. Isaiah (called "Osa" or "Ozzie"), 15, and John, 11, remained.

Bessie was 6 when she started school. Two years before, in 1896, the U.S. Supreme Court in the case of *Pessy v. Ferguson* had established the legality of "separate but equal" schools. Her separate-but-equal school was a one-room wooden building on Aiken Street in the already segregated black residential district of Waxahachie. Hot in the summer and cold in the winter, the room held students in grades one through eight. There was one teacher for all, one who probably lacked minimum qualifications for teaching. (As late as 1922 a survey by a professor of rural education at Columbia University claimed there were 15,000 black teachers in rural schools with no more than a sixth-grade education.)

Bessie walked four miles from her home to school, where she was taught reading, writing, and arithmetic, often without textbook or enough paper and pencils. Intelligent, uninhibited, and eager to learn, she quickly established herself as the star pupil in math.

In 1901 Bessie's life—until then a relatively happy one with ample time for play between school and chores—underwent a dramatic upheaval. George Coleman left his family. As a day laborer, he was denied the hoped-for prosperity promised by Waxahachie's booming cotton mills and warehouses. Being black, he was barred from voting by discriminatory poll taxes and "literary" tests. He was denied representation in state or local government and participation in land-grant programs. Recently passed Jim Crow laws forbade his riding in railroad cars with whites. He was denied the use of public facilities used by whites—restrooms, restaurants, hotel rooms, and even water fountains. If he protested he risked being whipped, tarred and feathered, or even lynched.

He had had, he told Susan, enough of Texas. His decision could have been prompted by news from Brenham, seventy miles northwest of Houston, that a mob of whites had rioted for two days to protest the employment of a black brakeman by a railroad company. Or by the fact that there had been 115 lynchings in the South the previous year.

George could not escape the bondage of race in Texas. Not only was he black but he came from a lineage of three Indian grandparents. Ironically, in Texas his Indian blood put him in even more jeopardy than did his being black. How most white Texans of the time viewed Indians is reflected in a history published in 1894 describing Ellis County Indians as "savage and treacherous" and predicting it was "doubtful whether one of their race will be living one hundred years hence."

In Oklahoma, however, that same Indian lineage could offer George an escape from the double bondage in Texas. If the family moved to Indian territory in Oklahoma, he told Susan, they would enjoy the full rights of citizens. He could go if he wanted to, Susan replied, but she was neither pioneer nor squaw. She and her children would remain in Waxahachie.

Soon after George left, John and Isaiah also departed. John, by then 15, joined his older brother Walter in Chicago where the latter worked as a Pullman porter on the Chicago–St. Paul run. Isaiah went to Canada where he became a successful farmer in the Amber Valley, 100 miles from Edmonton.

At 45, Susan Coleman was left with four small girls, the oldest age 9, at a time when 85 percent of black households were headed by males and single mothers were a decided minority. She had no kinfolk to help her as did most African American families in the South. Proud, inflexible, a loner sustained by her faith in the Lord and herself, Susan was on her own.

Within days of George's leaving she found work as a cook-housekeeper for a white couple, Mr. and Mrs. Elwin Jones. Considering how most black domestics were treated in Texas at the turn of the century, the Joneses were generous employers, willing to hire someone who could not "live in" and thus could not be on call twenty-four hours a day.

Mrs. Jones provided most of the food consumed by Susan and her children, sacks of flour and meat, and on bread-baking days told Susan to "bake more so you'll have enough to take home." The plants on the Coleman porch were given her by Mrs. Jones, who also clothed the Coleman girls with hand-me-downs from her own daughters' wardrobes. Some of these dresses had hardly been worn, given to Bessie and her siblings because they had admired them.

While her mother worked at the Jones house, Bessie took over as surrogate mother and housekeeper at the Coleman home on Mustang Creek. She washed, cooked, ironed, and cleaned, all without running water, electricity, or plumbing. Water was drawn from an outdoor well. Laundry was done in an iron tub, and meals prepared on a wood-burning stove that also heated the house.

Seven-year-old Elois noticed and recalled years later that "Bessie had little time to be a carefree child. She seldom took an interest in dolls though she would watch us play at times." There were days, even weeks, when Bessie missed school because she was needed at home to look after Nilus and Georgia. Only when these two reached school age were all four girls able to make the daily four-mile trek to school.

Apart from the rudimentary instruction in the "three Rs" she received from her one-room-school teacher at Aiken Street, Bessie's principal training and instruction came from her proud, hardworking, deeply religious mother. Susan made attendance at the Missionary Baptist Church mandatory for all the children. As soon as Bessie learned to read she was assigned a reading from the Bible every night after dinner.

"Even though Mother had to pinch pennies," Elois recalled, "she managed to get books from a wagon library that passed the house once or twice a year, telling of the accomplishments made by members of our race . . . Of course we learned Harriet Tubman at Mother's knees."

Susan also had hopes of her children achieving worldly success. For this she advocated their emulating educated white people like the Joneses. She observed the manners and speech of her employers, then brought her observations home to her children. Bessie and her sisters ate at a table covered with a cloth, and they ate with knife, fork, and spoon, just as the Joneses did, with many a "please," "thank you," and "excuse me," under the watchful eye of their mother.

Every year Bessie's carefully set routine of school, household chores, and church attendance was shattered by the cotton harvest. While cotton was king, Waxahachie's black schoolchildren clearly were his subjects and school shut down as soon as cotton picking began. Depending on the weather, this was usually late July or August. And no matter how long the harvest season might last, school stayed closed until all the cotton was picked, meaning that sometimes the fall school semester was delayed until November or even December.

It was a mutually beneficial arrangement. The cotton growers of Waxahachie needed every hand—man, woman, and child—and the Colemans and their neighbors needed the money. From the time the plants dropped their pink and white flowers and the cotton bolls showed their white fruit, Susan and her girls were in the fields. Excused from her work at the Joneses, Susan dragged a long bag, fastened by a strap around her neck, along each row, picking and putting the cotton into the bag until she reached the end of the row where she emptied it into a basket.

Until they left home Isaiah and John joined her and as each of the girls became strong enough to drag a sack they worked with their mother as pickers. The younger ones tagged along, fetching water or running errands for the harvesters. Backs bent and fingers bleeding, adults averaged between 150 and 200 pounds of cotton a day. One picker recalled, "Every colored person worked from the time he or she was old enough to drag a sack through the cotton fields. The work was back-breaking, exhausting and sometimes degrading."

At 10 Bessie was a reluctant, evasive picker. The bright, eager scholar who secretly dreamed of a better life she was not yet able to define showed her aversion to the debasing labor by deliberately lagging behind. On one occasion she was even caught riding on the sack of the picker in front of her.

Susan was aware of Bessie's errant behavior but was forced to overlook it because Bessie was the family accountant, the only Coleman who could record the weight of each sack accurately and make sure that the foreman totaled them correctly. Better yet, she dared to increase that total and add to the day's earnings by pressing a discreet foot on the platform of the weighing scale whenever the foreman happened to be looking the other way.

There was one break during the cotton-picking season—the day the circus came to town. Elois remembered how "we laid our cotton sacks aside and took a day off to go to the Ringling Brothers Circus with a shiny half dollar each to spend as we wished, for pink lemonade, and popcorn, balloons . . . . "

The circus came to Waxahachie only once a year. For the rest of the time Bessie was the family's chief source of entertainment. She would read aloud every night by the light of an oil lamp from the books that Susan rented from the "wagon library." Imitating the mellifluous voices of the black Baptist preachers she heard on Sundays, clever and talented Bessie lent added drama to the stories about "the Race." She read first from the Bible, at Susan's insistence, then the books she herself loved best, stories of African American heroes—Booker T. Washington, Paul Lawrence Dunbar, and Harriet Tubman among them. Her favorite book was fiction, *Uncle Tom's Cabin*, although she obviously thought that Tom and Topsy were feckless cowards. The night she finished reading the book for the first time she announced to her spellbound audience, "*I'll* never be a Topsy *or* an Uncle Tom!"

Young Bessie displayed an impressive self-confidence despite living in a world that seemed to demand her abasement as both a black and a female. To some degree her sense of worth was the result of being born healthy, intelligent, and beautiful. Some stemmed from her successfully performing the work of an adult even though still herself a child, looking after a home and three siblings. And some came from her certainty that since she had mastered reading and math there was nothing in the world she could not learn.

Underlying all of this was the example set by her mother whose own sense of worth was rooted in faith. To Susan's mind God created all people equal. So while rendering necessary respect to the white Caesars she never felt less than equal in the sight of God.

At 12, Bessie was formally accepted into the Missionary Baptist Church. (Later, as an adult and free from her mother's authority, Bessie attended church service less frequently, not because she had lost her basic belief but because God's heaven in the hereafter was not as immediately intriguing as the world around her.) In Waxahachie, however, she became an enthusiastic church fundraiser, and when a hand organ was offered as a prize to the person selling the most raffle tickets, it was Bessie who won.

Bessie completed all eight grades of Waxahachie's one-room black school. Limited and sporadic though her education was, it fed her growing hunger for more. And whatever that inchoate more was, Bessie suspected it was not to be found in Waxahachie. She had, in fact, been certain of this for some time and, with Susan's strong encouragement, began working as a laundress to earn money for whatever was to come.

In 1910 she took her savings and left for Langston, Oklahoma, to enroll in the Colored Agricultural and Normal University as "Elizabeth Coleman," a more elegant name than the one first given the census taker by her parents. The town of Langston, forty miles northeast of Oklahoma City, was founded in 1891. Named after John Mercer Langston, uncle to poet Langston Hughes's mother, it was the first of several all-black municipalities in Oklahoma. Seven years later the university opened. Belying its title, Langston, a state land-grant institution, was more a vocational school, offering four-year courses in education, agriculture, health services (home economics), and mechanical arts. Under the leadership of Dr. Inman Paige, still president when Bessie enrolled, the university also included a preparatory school for new students who fell short of full entrance requirements. There, at the age of 18, Bessie was placed in the sixth grade.

Bessie completed only one term before her money ran out and she was forced to return to Waxahachie. Whatever personal disappointment she must have suffered she kept to herself. Informed that the Missionary Baptist Church was planning to give her a homecoming party, she brought the entire Langston band home with her to provide the music. The little girl champion raffle ticket seller had matured into a young woman with a celebrity's talent for self-aggrandizement. She came home the conquering hero.

This brief moment of glory over, Bessie was back where she had started in east Waxahachie. If she had chosen to marry, as did most black women of her age and circumstances, her husband most likely would have been a tenant farmer or a cotton factory laborer. But she did not marry and resumed her work as a laundress.

As did most others like her Bessie collected and delivered the dirty laundry of her clients once a week, walking roughly five miles past the Romanesque-revival Ellis County Court House, whose pink and red limestone is still a landmark attraction, into west Waxahachie. She worked at home where she boiled the clothing in a tub in her backyard, scrubbed it on a washboard, rinsed, starched, and wrung it out, and hung it on clotheslines to dry. She ironed with a heavy iron heated on the top of the stove. On Saturdays she delivered her work, "keeping her place" by bringing it to the back doors of the west-side mansions. Elois said later that this and other humiliations failed to embitter Bessie who, like her mother Susan, never felt less than anyone else's equal.

While she worked Bessie dreamed of escaping Waxahachie. At night she read. Elois was often awakened at 2 or 3 o'clock in the morning for conversation that Bessie called "thinking." As Elois recalled, "Bessie would introduce a subject and I was supposed to respond."

"Thinking" was essential to the young woman so anomalous to the community in which she lived. Much of her "thinking" was stimulated by the brash exhortations of the African American weekly, the *Chicago Defender,* distributed by porters working on the southbound trains. The paper's advice was to "leave that benighted land [the South] . . . You are free men," it adjured. "To die from the bite of the frost is far more glorious than that of the mob." There were, it was claimed, high wages in the great terminus of the north-south railroads—as much as eight dollars a day.

In 1912 Bessie made less than that in a week. In Dallas a rented room was $2.00 a week, a hotel room with a telephone $1.00 a night, a silk dress $5.45, a leather purse $1.35, a Ford automobile $590, and an eight-room house with two baths $7,500.

There are no letters to prove it but Bessie, ever the opportunist, must have written to her brother Walter for help. In 1915, when she was twenty-three, Walter told her she could come to Chicago and stay at his apartment while she looked for work. Here, at last, was the chance for her to "amount to something," a goal Susan aspired to for all her children. "You can't make a race horse out of a mule," her mother told Bessie. "If you stay a mule, you'll never win the race." Bessie was certain that Chicago was the right track. She was going to be the race horse who came in first, the one that would "amount to something."

CHAPTER 2

That Wonderful Town

In 1915 Bessie Coleman left Waxahachie for Chicago. She took the Rock Island train after buying her ticket on the "colored side" of the depot off South Roger Street. Like most of her fellow pilgrims seeking the promised land of Chicago she was dressed in her "Sunday best," carrying all of her possessions in suitcases and bundles along with a sack of food for the twenty-hour ride ahead.

In those days blacks could travel in only two sections of a train, the all-black car or the Jim Crow section of a men's smoker. The former had a single toilet that would become unbearably filthy by the end of the trip. The Jim Crow section on the back of the smoker was separated from the white male passengers by a shoulder-high partition. A blue fog of smoke clung to the ceiling, the odor of ashes permeating the stale air.

The car on which Bessie sat on a hard wooden bench exemplified the oppression she longed to escape, a world of rear seats on buses and balcony seats in theaters, of forbidden public restaurants, water fountains, and lavatories, a world in which the old taboos and fears were now being augmented by a new one, the

15

resurrection of the Ku Klux Klan, first in the South and soon after throughout the nation.

Even as Bessie rode north, movie director D. W. Griffith's epic film *Birth of a Nation* was playing to packed theaters throughout the nation. Glorifying the anti-black, anti-Catholic, anti-Jewish Klan, the film was applauded by the white Protestants of Chicago. Its message, somewhat disguised in a coating of compassion for a South humiliated by its defeat in the Civil War, was that *only* white Protestants were genuine Americans, fit to run the country. Except for this elite in the big, ever-expanding metropolis of Chicago, ruled by millionaires and the political bosses they controlled, the Klan was a threat not only to blacks but to many of the city's immigrants. They were Germans, Poles, Irish, Hungarians, Russians, Greeks, Italians, and Slavs, mostly either Catholic or Jewish. Each national group formed a cultural enclave with its own schools, churches, synagogues, associations, and clubs. In time there was assimilation, but not for African Americans, whose skin color proved an effective barrier. Shunned by all the other groups, blacks erected their own enclave. Pride, habit, and the need for mutual protection led to the establishment of a self-segregated ghetto with its own churches, clubs, and fraternal organizations.

Fifteen years after Bessie arrived in Chicago the attitude of most whites throughout the nation toward African Americans was still so demeaning that a reputable New York firm published a book by a Chicago historian describing the typical black as "a virtual savage from the cotton fields." It went on to say that although he might "don clothes and something of the manner of an urban dweller, . . . the Negro, half-child, half-man, was still the under dog. He was still a 'separate' being . . . He threw out his chest and jingled his silver, to keep up his spirits. He was delighted when white people treated him nicely."

Bessie was leaving the segregation of the South for the ghetto of the North. But within its borders, black Chicago offered the excitement and opportunities of big city life.

No one knows how Bessie made her way from the South Side station to her brother Walter's apartment at 3315 Forest (now Giles) Avenue, but she did and was welcomed. The first of George and Bessie Coleman's children to join the northern exodus, Walter had lived in Chicago for more than a decade. A gentle, quiet man of 35 who was called "Bud" by his siblings and friends, he was married to an acid-tongued woman named Willie from Tyler, Texas. Living with them was brother John, six years Walter's junior, and John's wife, Elizabeth. Both couples were childless.

Walter was a Pullman porter, a respected position within the black community but on the road a difficult and demanding job in which his race was a constant burden. On the train he washed in a separate washroom, drank from a separate fountain and slept, when he slept at all, on inferior blue sheets because the white sheets were "for white bodies only." The rules for porters were issued by the president of the Pullman Company, Robert Todd Lincoln, son of the late president.

Bessie quickly recognized that Walter was the main provider for the crowded Coleman household. John, a charming extrovert given to drinking bouts, was frequently unemployed. And it was Walter who had found the apartment and paid the rent. The Coleman housing situation was typical of the Chicago of their time, confined to the "Black Belt" by racial and social pressures and limited in their choices by the incredible influx of Southern migrants, who doubled the city's black population between 1910 and 1920. By the end of that decade, in fact, 90 percent of the African American population of the city would be crammed into housing in an area bordered by Twelfth and Thirty-ninth streets on the north and south and Lake Michigan and Wentworth Avenue on the east and west. Unlike the urban ghetto it has since become with its disproportionate number of impoverished, undereducated, and underemployed, in Bessie's time the area consisted of a number of mixed and balanced communities in which the wealthy and the well-educated, the large middle-class and the poor and the hard-working coexisted in generally law-abiding harmony.

Bessie clashed frequently with her sister-in-law Willie, whom she considered "too bossy" and, she said, dominated everyone in the household—everyone, that is, except Bessie. But she continued to live there because she loved Chicago and was determined to stay.

From almost the moment she arrived Bessie started looking for a job. In 1915 most black women who worked outside the home were domestics, although a few escaped that fate by becoming school-teachers. Not until the First World War would factory jobs open up for black women. But Bessie had not left Texas to be a maid, cook, or laundress. After a lengthy search she decided to be a beautician. Though perhaps somewhat comparable to domestic service in that she would be providing personal services, at least it would keep her out of the kitchen. Best of all, there were plenty of openings available in the numerous beauty shops of Chicago's South Side.

The thriving profession Bessie chose was the target of considerable criticism from African Americans who considered much of it simply a hypocritical attempt by blacks to imitate and emulate a ruling white society. One such critic, noted educator Nannie Helen Burroughs, wrote, "What every woman who bleaches and straightens out needs is not her appearance changed, but her mind changed . . . If Negro women would use half the time they spend on trying to get white, to get better, the race would move forward a pace."

Although the black press overflowed with similar exhortations to take pride in "the Race," many advertisements were for skin lighteners and hair straighteners. Two-column ads urged readers to use Nadinola Bleaching Cream ("See how it lightens and whitens your skin overnight"). Others advocated using Pluko White hair dressing, Arroway hair cream for men, and caps for making hair straight and smooth. The firm of Poro Products for hair and skin frequently offered to train sales personnel. Most prominent and plentiful of all were the ubiquitous solicitations of saleswoman supreme Madame C. J. Walker whose "wonderful hair and skin preparations" had already made her one of the wealthiest woman entrepreneurs in the country—black or white.

The debate continued into almost the next decade with sociologist Frances Marion Dunford warning in the *Journal of Social Forces* that the penchant of the black male for "using materials which tend to make him more like his white brother" would be the downfall of the race. An outraged *Chicago Defender* writer's response was that black people follow fashion just as most people in any society do. The progress of the race, he wrote, would be determined by its ability to assimilate within the environment of which it is a part. Segregation would remove the race from the environment, making assimilation, and progress for the race, impossible.

Bessie was untroubled by the debate that swirled around her. She was going to be a beautician because it was the most interesting and appealing employment she could find. Her own appearance changed as she experimented with makeup, hairstyles, and clothing. Photographs and the recollections of her two nieces, Marion and Vera, indicate that her choices were eclectic. She used whatever was most becoming to her and seemed most to enhance her own figure, face, and personality.

Bessie enrolled in the Burnham School of Beauty Culture for a course in manicuring—or so she wrote to Elois, who was still in Texas. Chicago's city directory of 1915 identifies Edward Burnham as a distributor of hair-care goods with offices on North State Street and West Washington Street. But there is no Burnham School listed. Regardless of where Bessie took her apprenticeship, she did learn the beautician's trade. A year later she sent Elois a clipping from the *Chicago Defender* describing her as the winner of a contest to discover the best and fastest manicurist in black Chicago.

She certainly was one of the shrewdest. By confining her work to manicures she avoided the time-consuming lessons that would have been required of her before she could work as a hairdresser. As a manicurist she could do men's nails in a barbershop, where the customers appreciated her looks and charm and expressed their admiration in generous tips. She worked first in a shop at 3447 State Street and later in a barbershop owned by John M. Duncan at 206 East Thirty-sixth Street. At Duncan's she sat at a table in the window where her customers could enjoy being seen having their nails done by a very pretty woman.

The shop on State Street where Bessie first worked was in the area known as "The Stroll," eight blocks of State between Thirty-first and Thirty-ninth streets. The eight blocks were best described by Chicago historian Dempsey Travis as a "black Wall Street and Broadway." It was a place in which to promenade, to make bets, to sell goods, legal or illegal, to start a business, to talk politics, to bank, to go dancing, to watch a show, and to see and be seen. It was downtown, but with an ethnic identity. On the sidewalks people in colorful, often expensive clothing greeted one another warmly or gathered on corners for exuberant, sometimes noisy conversation. Along with obvious communal friendliness and uninhibited laughter there was, for many, a firm belief in luck, made evident by the sale of "dream books." These offered guidance to picking a number in an illegal lottery, the "numbers game." Sales of some—"The Three Witches," "Gypsy Witch," "Japanese Fate" and "Aunt Della"—were topped only by sales of the Holy Bible.

Bessie knew every inch of this territory and explored it all. She went to the dozens of nightclubs squeezed among banks, bars, shops, and restaurants along the Stroll. She worked next door to the Elite Club #2 owned by Teenan Jones. Jones had opened the original Elite Club in Hyde Park at Fifty-sixth and Lake streets but had been forced to close it by anti-black neighborhood groups in

1910. He moved it to State Street and added the second club to compete with other popular spots such as the Royal Garden, Pekin Café, and Dreamland. Bessie went to all of them and saw, among the great black performers of the day, "King" Joe Oliver, Louis Armstrong, Bessie Smith, Ethel Waters, and Alberta Hunter.

These nightspots were open to both races. Dancing was uninhibited and, even after the Eighteenth Amendment banned its sale in 1919, liquor flowed freely. Typical was the Pekin Café owned by undertaker Dan Jackson, where a large amount of illegal whiskey was discovered after two white detectives from Detroit were found murdered on the premises. The African American *Birmingham Reporter* called the Pekin "a Chicago rowdy house where whites and blacks meet." The same newspaper noted that Dreamland was temporarily closed soon after during one of Chicago's sporadic "rigid law and order movements" to clean up the "Great Light Way" where "amusements and other things" were offered.

Bessie's brother John was well acquainted with the patrons, both black and white, of places such as Dreamland. He worked as a cook for several bootleggers and gangsters. Bessie also knew them through her work. Men who wanted their appearance to reflect their status spent lavishly on their attire and grooming and came to Bessie at the shop on State Street for their manicures.

But not all of Bessie's continuing education was gained in the shop or on the Stroll. She remained an avid reader, her favorite source of information now being the *Chicago Defender*; its editor and publisher, Robert Abbott, became her idol. Spokesman for the race and owner of a newspaper whose readers would soon number a half million, Robert Sengstacke Abbott was a handsome, elegantly dressed, still-youthful man in his mid forties. He would stand on the corner of Thirty-fifth and State, cigar clamped between his teeth, and hold court as he chatted with community leaders. Abbott told them what they ought to do and they often did it. Bessie knew all about him and had seen and heard him often before the two actually met.

At that same corner of the Stroll Bessie mingled with others of the rich and powerful in black Chicago, among them Jesse Binga, whom she would later know intimately and who, along with Abbott, would influence her career. The notorious Binga, an ex-barber and Pullman porter, began his fortune in 1907 as a "block-buster" real estate dealer. He would buy one house on a white block and sell it to a black family, then buy up the remaining houses at lower and lower prices as the white owners rushed to sell them cheaply and move out.

Unloved, but respected and even feared by his own people, Binga also had white enemies who bombed his $30,000 house in the Englewood district seven times in less than two years. "As usual," one report said, "the pillars of the front porch were blown out of place and scores of window panes in the neighborhood shattered. Binga and his family were out of the city and the only one in the home was a maid who locked herself in and refused to open the door."

Binga took his money to the bank in a Model-T truck with one man driving and another armed with a sawed-off shotgun. "He was a mean son-of-a-bitch," one of his employees said later, adding nonetheless that invitations to his "high-class" parties were coveted by many. By 1925 Binga had a million-dollar bank and office building at the corner of Thirty-fifth and State. A reporter for a Baltimore black weekly wrote, "Walk down State Street and you will find two highly illuminated colored banks, with electric signs sufficient to attract you a mile away. On the signs you will read these words in color: Douglas National Bank. Binga State Bank."

Binga frequently met another of Bessie's acquaintances, his friend Oscar DePriest, Chicago's only black alderman and, later, U.S. congressman. DePriest, who also dealt in real estate, was tried but acquitted in 1917 on charges of corruption. Chief witness against him was nightclub owner Teenan Jones, who claimed he was DePriest's bagman and head of a gambling combine.

In addition to Binga and DePriest, Bessie knew Dan Jackson,

lawyer Edward Wright, and Anthony Overton, a cosmetics manufacturer and, in time, founder of a bank that would fail when all of
its capital was found to have been invested in one of his own
businesses.

For more than three years while she manicured nails and added
to her circle of friends Bessie showed no particular interest in any
one man. Then, on January 30, 1917, four days after her twenty-
fifth birthday, the manicurist who loved the bright lights of the
Stroll, the woman who had spoke so often of her determination to
"amount to something," married Claude Glenn, a friend of her
brother Walter's and fourteen years her senior.

The marriage must have astonished the members of her family—
if they knew about it at all. Sister Elois never mentioned it in her
memoirs of Bessie's life. When queried some seventy years later,
none of Bessie's four nieces and nephews knew anything about it.
The two nieces who had lived with Bessie in Chicago did recall that
"Uncle Claude," a tall, dark-skinned man, had been a caller at both
Susan's and Bessie's apartments. But they never suspected he was
anything more than a family friend. Nor did their Aunt Bessie ever
tell them she was Glenn's wife.

The only real evidence of the marriage is a license issued on
December 30, 1916, by Cook County Clerk Robert M. Switzer.
Added to the bottom of the license is a record of the marriage
ceremony performed by Baptist minister John F. Thomas at 3629
Vernon Avenue—probably the minister's home—a few blocks
from Walter's apartment. So far as is known, Glenn never lived at
the same residence as Bessie. Soon after the marriage Bessie
moved from Walter's apartment to one of her own at 3935 Forest
Avenue. But the city directory never listed Claude Glenn at that
address—not that year or later.

The circumstances surrounding Bessie's marriage are unclear.
Whatever her reasons, Bessie certainly didn't marry Glenn for his
money; he was no wealthier than Walter. Glenn was neither handsome nor dynamic, just a pleasant, quiet, older man. Nor had

Bessie ever shown any evidence of sexual attraction for him during his frequent calls at her apartment or Susan's. It is possible that Bessie found in him a dear friend when she needed one and that she married believing that friendship was all she needed.

During the next year it became increasingly clear that for Bessie the vow that took precedence over all others was the one she had made as a child to "amount to something," to be noticed and respected by as wide a circle of admirers as possible. Marriage, especially in that era, required far more subservience than she could bring herself to allow.

Bessie's mother and three sisters were equally strong-willed, all abandoning relationships they found distasteful or unbearable. Soon after Bessie's marriage Susan arrived in Chicago with her youngest daughter, Georgia, and Georgia's one-year-old daughter, Marion. Georgia may not have been married. When asked by Marion years later her mother refused to say whether or not she had wed Marion's father. Her only comment was, "I couldn't stand him."

In 1918 Elois was the next to come to Chicago from Texas. She arrived with four children, Eulah B., Vera, Julius, and Dean, fleeing the beatings of her cruel, hot-tempered spouse, Lyle Burnett Stallworth. Not long after Elois, the last sister, Nilus, came to the city with her four-year-old son, Arthur, leaving her husband, Willy Freeman, in Oklahoma. All three sisters, their children, and their mother, Susan, stayed briefly with either Walter or Bessie until they could find places of their own.

Three months after Bessie's marriage the United States declared war on Germany. Both Walter and John were members of the all-black Eighth Army National Guard. Although both men ultimately were sent to France, Bessie was more worried about attacks on them by white racists than she was about their becoming casualties at the hands of the Germans. Segregated and often mistreated by an Army command that used them as stevedores, black men in uniform were perceived by many whites as threatening. Those who

regarded African Americans as inferior refused, often using violence, to acknowledge the equality the uniform seemed to demand.
In Houston that summer twelve civilians died in a riot that started
when black soldiers attempted to board a streetcar reserved for
whites. In the aftermath, thirteen soldiers were sentenced to death
and another fourteen to life imprisonment. In Spartanburg, South
Carolina, black musician Noble Sissle, in Army uniform, was
struck repeatedly by whites after he failed to remove his cap in a
hotel lobby. The Army forestalled retaliation by hastily issuing a
directive ordering the black unit to break camp and leave for
France immediately.

John and Walter both served in France and survived the war
unharmed. Their outfit, the Eighth Army National Guard, was part
of the 370th Infantry, whose men had won twenty-one American
Distinguished Service Crosses and sixty-eight French War Crosses.
When the 370th arrived in Chicago, the entire regiment paraded
down Michigan Avenue in full uniform, but received little or no
applause from Chicago's white onlookers. Returning white veterans, many from ethnic groups whose economic and social status
was just one step above the African Americans, discovered their
jobs had been taken by blacks. Twenty percent of the meat-
packing force was now black. In eighteen months the city's black
population had increased from 50,000 to 150,000. Some of the
migrants displaced men who had enlisted or been drafted into the
armed forces. Others were needed for new jobs created by the
increased production demands of the war. Still others had been
hired as "scabs" to replace striking workers.

In March of 1919 the re-election of Chicago's mayor William
Hale Thompson was made possible by the votes of the city's two
black wards on the South Side. "Big Bill," six feet, two inches tall
and 260 pounds wide, had wooed and won black voters with rousing speeches in which he called them "brothers" and "sisters" and
kissed their babies. Yet he did nothing to abate the rampant racial
hatred aroused in jobless whites caught in a postwar depression.

Four months later Chicago was battered by the worst race riot in its history, touched off on July 27, 1919, when a black youth on a homemade raft drifted into an area of Lake Michigan customarily used only by whites. He was stoned by a group of white bathers, fell into the water, and drowned. That night groups of blacks and whites fought in the street with guns and knives. The violence carried over into the next day when gangs of whites, many of them unemployed, dragged blacks off streetcar platforms and beat them.

Just a block from Bessie's apartment 10-year-old Joe Crawford found shelter with relatives after fleeing from the violence in his racially mixed neighborhood. A white woman on the second floor of his apartment building had warned Joe's mother that a group of teenagers planned to give her and her son a public flogging. As Joe and his mother fled out the back they heard voices at the front door shouting, "Niggers, come out and get your asses whipped or stay in there and be barbecued!"

Four days later the National Guard, aided by a torrential rainstorm, restored order to the city. The riots had left 38 dead, 537 injured, and more than 1,000 people homeless. Bessie and her family all escaped injury or property loss.

At the close of that summer Bessie had been in Chicago almost five years. She had moved north, settled in Chicago, learned a trade, married, found her own place to live, seen her brothers go to war, and survived a race riot. But at 27 she was still looking for a way to "amount to something."

The answer came unexpectedly one day that fall when she was working at her manicurist's table in Duncan's barbershop. Her brother John, who had failed to "settle down" after his wartime service, walked in. Far from sober, he began a teasing discourse— one he had given before—aimed at Bessie. His theme was the superiority of French women over those of Chicago's South Side. French women, John said, had careers. They even flew airplanes. "You nigger women ain't never goin' to fly," he told Bessie. "Not like those women I saw in France."

After the laughter from his captive audience of barbershop customers had subsided, John looked over at Bessie. She was smiling at him. "That's it!" she said. "You just called it for me."

John's comment no doubt rankled, but Bessie's response was clearly more than just a spur-of-the-moment reaction. During the war Bessie had read the stories and seen the photographs of aviation heroes. And John had talked about French women fliers before. Clearly Bessie had decided that flying would be her new vocation. The air would be the arena for her ambitions, a way for her to be noticed.

CHAPTER 3

Mlle. Bessie Coleman—
Pilote Aviateur

From the moment Bessie decided to become a pilot nothing deterred her. The respect and attention she longed for, her need to "amount to something," were directed at last toward a definite goal. Ignoring all the difficulties of her sex and race, her limited schooling and present occupation, she set off to find a teacher. She approached a number of fliers—all of them white, since there were no black aviators in the area at the time—but they all refused.

Perhaps they did so because her good looks and self-assurance in the presence of whites struck them as brash and suggested to them more a publicity seeker than a dedicated student. Perhaps they turned her down because she was black. Or because she was a woman. Most likely it was a mix of all three with race and gender ranked first and second.

At any rate, Bessie was forced back to the South Side, to her own community and to the man she so admired, Robert Abbott, editor-publisher of the *Chicago Defender*. She wanted to become a pilot, she told him, but no one would teach her. What should she do?

29

Abbott saw nothing outrageous in Bessie's objective. The paper he had established in 1905 aimed to gain recognition for black Americans as a people worthy of respect by all Americans. Unlike many of his gender, black or white, Abbott didn't think of women as less capable than men. His background wouldn't let him. His aunt, Priscilla Hammond, was a founding member of St. Augustine's Episcopal Church in Atlanta, Georgia. Another aunt, Cecilia Abbott, founded St. Stephen's Episcopal, also in Atlanta. His cousin Roberta Gwendolyn Thomas was a student at American University in Washington, D.C. And when his mother, Flora Abbott Sengstacke, came visiting from Atlanta, she was met by no less than Julius Rosenwald, president of Sears Roebuck and noted benefactor of black schools.

Abbott said Bessie must go to France. The French, he claimed (as did his frequent newspaper editorials), were no racists. They were also leaders of the world in aviation. If she worked hard, saved her money, and learned French, Abbott told her, he would inquire about an accredited aviation school in France and provide her with a reference.

Taking Abbott's advice Bessie enrolled in a language school on Michigan Avenue. She found a better-paying job as manager of a chili parlor at Thirty-fifth Street and Indiana Avenue and began to save her money, putting it in the Franklin Trust and Savings Bank at 100 East Thirty-fifth. Sister Georgia took over her manicurist's table at Duncan's after Bessie trained her. With Georgia now earning money and contributing to the rent of the apartment she shared with their mother at 3757 Indiana, Bessie moved a few blocks up to a new apartment of her own at 4533 Indiana.

It is difficult to believe that Bessie's wages and tips as a chili-parlor manager were enough to pay for passage to Europe and back, as well as room, board, and tuition for the flying lessons. Presumably Jessie Binga, who was already a rich man when Bessie was still a schoolgirl, gave much of the money. Gossip reputed her to be Binga's mistress although there is no real evidence of this.

Robert Abbott's interest was platonic, but he did give her money as well as guidance in his desire to see her become a pilot. That interest arose, one of his reporters said, from his tireless pursuit of increased readership for the *Defender* rather than any real concern for Bessie. As the first African American woman pilot, she would provide a pride-in-race increase in circulation for his newspaper.

There are other possible sources of the funds she needed. In an interview several years later Bessie mentioned an unnamed Spaniard who "made it possible for me to continue my studies in aviation." "She had a lot of men callers," her niece Marion recalled. "Some were black and others were white—and of several nationalities. I remember hearing different languages." While there is no evidence that any money changed hands or that any of these men received sexual favors for their assistance in financing Bessie Coleman's career in aviation, it is clear that Bessie lived as she pleased. The understandings of the times are perhaps the best guide.

On November 4, 1920, Bessie applied for an American passport at the Chicago office. By then, with Abbott's help, she had located an accredited aviation school in France and had learned enough French to read most of the school's first reply. However, her writing was no better than a child's and she left the second "t" out of Atlanta.

She put January 20, 1896, as her birthdate, four years less than her age. On this sworn statement she gave her occupation as manicurist and stated she had never been married. The purpose of her visit to "England, France and Italy" was "to study."

On the application, John Coleman also gave 4533 Indiana Avenue as his address. As a character witness he swore that his sister was an American citizen, born in Atlanta, Texas, on the date she had given. Further identification was provided by a second witness, Mrs. Anna M. Tyson of the same address who gave her occupation as "housekeeper."

The deputy clerk of the U.S. District Court wrote the description of the applicant as "twenty-four; five feet, three and half inches in height; a high forehead; brown skin; brown eyes; a sharp nose and medium mouth; a round chin and brown hair." The brown skin was copper-colored and "sharp" defined a nose more Caucasian than African. The attached photograph is of a very pretty woman.

A week after the passport was issued on November 9, Bessie went to the British and French consulates in Chicago for two visas. The first, from the British, was a month's transit permit; the second, from the French, was a tourist's visa for a year, valid until November 16, 1921. On November 20 Bessie sailed for France from New York City on the S.S. *Imparator*.

In her account of the months she spent in France, Bessie said, "I first went to Paris and decided on a school. But the first to which I applied would not take women because two women had lost their lives at the game." Forced to find another school, Bessie looked for the best and found it "in the Some, Crotoy where Joan of Arc was held prisoner by the English." She had selected France's most famous flight school—École d'Aviation des Frères Caudron at Le Crotoy in the Somme—managed by French aviators and plane designers Gaston and René Caudron. There she completed what she said was a ten-month course (it was actually seven months), "including tail spins, banking and looping the loop."

During her training at Caudron Bessie witnessed a terrible accident in which another student pilot was killed. "It was a terrible shock to my nerves," she said, "but I never lost them; I kept going," although, she added, she had to sign away her life by agreeing to take all responsibility for her injury or death.

Bessie must have longed to see people of her own race, for she mentioned that only two non-Caucasian students were in the class and she did not identify them as black. Her room, she said, was nine miles from the airfield, nine miles that she walked "every day for ten months." Without family or friends, obliged to speak in a language she had not yet mastered, Bessie finished the course in June, her name appearing in the registry of 1921 graduates.

Bessie learned to fly in a French Nieuport Type 82 (School). Just as the Curtiss JN-4, or "Jenny," was a favorite learner's plane in the United States, the Nieuport was frequently used in France for teaching pilots, both before and after World War I. The twenty-seven-foot biplane with a forty-foot wing span was designed by Gustave Delage and manufactured by the Société Anonyme des Establishments Nieuport. It was a fragile vehicle of wood, wire, steel, aluminum, cloth, and pressed cardboard. Structural failure, often in the air, was all too common. Bessie probably heard the pilots joke about the cloth-covered wings that they claimed were made by an insane seamstress whose work tended to peel off the almost forty-foot span of the upper wing.

Each time she took a lesson in the Nieuport with its dual controls, she had to inspect the entire plane first for possible faults—wings, struts, wires, cloth covering, engine, propeller, cowling, and the four-wheel landing gear with pneumatic tires. The plane had no steering wheel or brakes. The steering system consisted of a vertical stick the thickness of a baseball bat in front of the pilot and a rudder bar under the pilot's feet. These were duplicated for the student in the rear cockpit. The stick controlled the pitch and roll of the aircraft. The rudder bar controlled its yaw, or vertical movement. To stop after landing, the pilot would lower the tail of the plane until a rigid metal tail skid dug into the earth.

Once in the cockpit, Bessie had to wait for a mechanic to start the engine, priming it first with castor oil. She could feel the heat from the eighty-horsepower engine and smell the overpowering odor of hot castor oil that covered her face, goggles, and leather coat with a fine yellow mist.

Bessie could not always see what the instructor was doing or hear his comments with the engine's roar drowning out all other sounds. She learned by watching her stick and rudder bar move as the pilot moved his. Soon she was placing her hands on the stick and her feet on the rudder bar, enabling her to get the feel of what the pilot did. This primitive dual-control system posed a risk to both Bessie and her instructor. If she were to "freeze" on the stick, to grasp it so tightly in a panic that the pilot could not regain control of it, both could die, a tragedy that occurred all too frequently for students and instructors at that time.

After seven months of lessons and practice flights Bessie took the test qualifying her for a license from the renowned Fédération Aéronautique Internationale (FAI), the only organization at the time whose recognition granted one the right to fly anywhere in the world. She flew a five-kilometer closed-circuit course twice, climbing to an altitude of fifty meters, negotiating a figure eight, landing within fifty meters of a predesignated point, and turning off the engine before touching down.

Bessie's coveted FAI license is dated June 15, 1921. The document accurately recorded her name (Bessie Coleman), her birthplace (Atlanta, Texas) and, somewhat less accurately, her age (25, the figure Bessie gave the passport authorities in Chicago, instead of the 29 that she really was). It did not record that Bessie was the first black woman ever to win a license from the prestigious FAI and of the sixty-two candidates to earn FAI licenses during that six-month period, Bessie was the *only* woman.

From Le Crotoy Bessie went to Paris in June to take more lessons from, she said, an unnamed French ace who had shot down thirty-one planes in the war. There is no record of Bessie's stay during the next two months but she arrived at a time when summer tourists filled the city's hotels and rooming houses. An acquaintance, Dr. Wilberforce Williams, who was also a columnist for the *Chicago Defender*, didn't reach the city until after Bessie had left but his impressions of France indicate an atmosphere in which Bessie must have reveled. "In France," he wrote, "more than any other country, one finds the privileges of individual freedom and political unity. There is a total absence of racial antagonism."

Aside from her flying lessons Bessie must have done what most tourists did—explored the city, sat at sidewalk cafés, walked a good deal, and perhaps saw some museums and landmarks. She certainly shopped, for she brought home a stunning wardrobe including dresses, a tailored flying suit, and a leather coat.

Bessie left France on September 16, 1921, on the S.S. *Manchuria* from Cherbourg to New York. This time reporters—black and white—met her in New York to interview her. The *Air Service News* reporter stated that this "twenty-four-year-old Negro woman" had returned "as a full-fledged aviatrix, the first of her race," and that she intended to give exhibition flights in the United States. His competitor from the *Aerial Age Weekly* noted that she arrived with credentials from the French "certifying that she has qualified as an aviatrix." Adding to the aviation magazines' stories the *New York Tribune* reported Bessie's claim that she had ordered a Nieuport scout plane to be built for her in France.

For most of the country's black weeklies Bessie was a front-page story. The *Dallas Express* quoted her as saying that few colored people and no women had taken any interest in aviation. Using India as an example, she said, "Out of four hundred million Hindus only one has piloted a plane, and that was a man."

Before leaving New York City for Chicago Bessie was the guest of honor at a performance of the musical *Shuffle Along*. Written and produced by two vaudeville comedians, Flourney Miller and Aubrey Lyles, the all-black musical was first presented in one-night stands with no more scenery than could be moved in one taxi. After every Broadway theater asked had turned it down, the show opened a mile from Broadway at the Sixty-third Street Music Hall, where a special stage had to be built for it. Yet, in a summer plagued by a heat wave, *Shuffle Along* was an instant hit, playing to standing-room-only audiences. The *Shuffle Along* cast included Eubie Blake, Ethel Waters, Florence Mills, Roger Matthews, Lottie Gee, and Noble Sissle, who recorded the music on the Emerson label. The women in the chorus were beautiful and talented. (One aspiring performer who didn't make it, however, was 15-year-old Josephine Baker, turned down because she was "too young and too dark.")

When Bessie appeared on stage to receive a silver cup engraved with the names of the cast the entire audience—whites in the 550 orchestra seats reserved for them and blacks in the balcony and boxes—rose to applaud the only black woman aviator in the world.

Returning to Chicago in October Bessie was interviewed in her apartment at 4533 Indiana Avenue by a reporter from the *Chicago Defender*. Because the *Defender* had used Bessie's picture the previous week, on this occasion they photographed a proud Susan holding the silver cup presented to her daughter by the *Shuffle Along* cast.

For the *Defender* interview Bessie gave a complete account of her flying experiences in France, beginning with her failure to find a teacher in Paris, then her move to Le Crotoy, the lessons and the risks. After passing her tests, she said, she returned to Paris for more lessons with the French ace. She liked to fly high, she said, because "the higher you fly, the better the chance you have in case of an accident," meaning she would have more time to correct a problem.

Asked why she wanted to fly, Bessie said, "We must have aviators if we are to keep up with the times. I shall never be satisfied until we have men of the Race who can fly. Do you know you have never lived until you have flown? Of course, it takes one with courage, nerve and ambition. But I am thrilled to know we have men who are physically fit; now what is needed is men who are not afraid of death." She offered to meet with and provide information on flying schools to anyone fearless enough to meet that challenge.

Bessie told the *Defender* reporter that she had ordered a plane built for her own use, a "Nieuport de Chasse" with a 130-horsepower engine, in which she intended to give exhibition flights "in America and other countries." During her stay in France she had seen, she said, "fine Goliath airplanes, the largest built by the house of Farman, equipped with two Sampson motors, planes shown only to flyers because, although they could carry up to fourteen passengers, they were fitted out as war planes."

The Goliaths, built by the Société des Avions H. & M. Farman, actually carried twelve passengers, a pilot, and a mechanic, and some indeed were fitted out as bombers. So Bessie was not too far off the mark. But the Nieuport she claimed to have ordered was never delivered.

Clearly Bessie was prone to exaggeration and not all of her stories were true. Still, she was no more guilty of enhancing her interviews with falsehoods than most of her pioneer aviator colleagues. She knew instinctively that in this postwar period of avid hero worship for cinema stars, athletes, and aviators, the more sensational the story, the better-known the storyteller. And, with radio in its infancy and television not yet on the horizon, there remained only the newspapers to report and relate their activities to a naive public of eager believers. That same public could not imagine aviation as anything other than wartime weaponry or high-risk amusement—the risk for the flier, the amusement for the observer. With regular commercial flights a decade or more in the future, the only way that most civilian aviators could earn a living was to give air shows for paying audiences. But it took publicity to get those audiences and for that Bessie and her like were almost totally dependent on the press. Only the print media—the daily and weekly newspapers—could create their public image and ensure their livelihood. Bessie, even more than most of her colleagues and competitors, quickly sensed that to succeed in this strange mixed world of business and entertainment and showmanship, a certain amount of embroidering the truth, of flamboyance and flair, was a virtual necessity. And for a black flier, a story *had* to be sensational to grab the attention of the white press.

But for all her embellishment of her own adventures Bessie's pride in her race was deep and genuine. When a reporter from the *Chicago Herald* offered to do a story on her if she agreed to pass as white, she took her mother and niece along with her for the interview. She was laughing as they walked into the reporter's office. Pointing to Susan and Marion, who were dark-skinned, she said, "This is my mother and this is my niece. And you want me to pass?"

Bessie soon realized that it was difficult for any pilot, let alone a black one, to earn a living flying. Most of them became "barnstormers" in shows called "flying circuses." These aviation pioneers moved from town to town, renting unused patches of farmland (and often sleeping in the farmers' barns) to gain paying audiences for their demonstrations of aerial acrobatics to the curious. In war-surplus planes of fabric and wood, they "looped the loop," deliberately stalled in midair, dived, and barrel-rolled at the risk of their lives.

While Bessie was still in France a woman flier, Laura Brownell, had set a "loop-the-loop" record for women of 199 loops at Mineola, New York. But most barnstormers were men, with women consigned to the role of wing-walker or parachutist. A month before Bessie returned from Europe, Phoebe Fairgrave, later a trophy-winning speed pilot, set a parachute-jump record for women of 15,200 feet. Fairgrave did the jumps to pay for flying instruction and later opened an air-show company of her own. Bessie was back in Chicago only ten days when a Chicago waitress, Lillian Boyer, made her first plane-to-plane transfer and within the next year developed a stunt in which she transferred to a plane from a speeding automobile by means of a rope ladder.

Feature stories in the *Defender* and admiration for her Paris wardrobe were not going to get Bessie a job flying. And her FAI certificate was of little use since licenses weren't required to fly in the United States. Nor did it take a genius to see that mere figure eights and pinpoint landings would never draw a paying audience looking for thrills and excitement. To join an air circus she would need more lessons to enhance her skills and her flying repertoire. There was still no one willing to teach her in Chicago. She would have to return to France.

Where the money for further lessons came from, and whether it came from one source or more, no one really knows. But in circumstances different from her first departure Bessie now had not only chic Paris gowns and attractive leather flying apparel but much favorable newspaper exposure, especially in hometown supporter Robert Abbott's *Chicago Defender*. In early February she booked passage on the S.S. *Paris* and left New York City on February 22 for more training.

CHAPTER 4

Second Time Around

Bessie arrived in New York a week before she was scheduled to sail on the S.S. *Paris* for Le Havre. It was a week in which she was much sought after and lionized by black New Yorkers. The *Shuffle Along* company had moved on to Boston but others in the vibrant community of Harlem had read her press notices and were eager to meet her. She was the guest of a woman she claimed was her aunt, a Mrs. Robinson, at 36 West 136th Street, a house on a site now occupied by the Harlem Medical Center, across the street from the New York Public Library's branch on Malcolm X Boulevard. This was the heart of Harlem, an area described by African American scholar James Weldon Johnson as "twenty-five blocks northward from 125th Street and Fifth Avenue, bordered on the east by Eighth Avenue and the west by St. Nicholas Avenue." No member of the Coleman family knew of an aunt named Robinson but she may have been an older woman whom Bessie referred to as "aunt" rather than "sister," a common reference in the black community.

Bessie had arrived in New York during the beginning of a creative explosion of African American art. Poets, novelists, artists, and musicians, later dubbed the "Niggerati" by anthropologist, novelist, and playwright Zora Neale Hurston, were all demanding justice and equality for their race. Among the most popular of these social revolutionaries were poets Claude McKay, Countee Cullen, and Langston Hughes, all of whom gave frequent poetry readings. Joining in their call for change was the commanding voice of Marcus Garvey, who advocated the establishment of a new African American nation in Africa.

Yet the new, free voices of Harlem were often ignored by the majority of whites. The white Drama League had voted black actor Charles Gilpin, star of Eugene O'Neill's *Emperor Jones,* one of the ten persons most advancing the American theater. But when Gilpin, then appearing in O'Neill's work at the Maryland Theater in Baltimore, tried to buy an orchestra ticket at the box office, the great star was refused. "Colored" were allowed balcony seats only.

While blacks were refused orchestra seats in white theaters, whites—and their money—were welcomed in Harlem's nightclubs. Anxious to test the new freedoms from prewar social restrictions, some whites used Harlem as their testing ground, believing that there they could behave as they imagined black people did, expressing themselves uninhibitedly in a manner that would not have been sanctioned in their own circles. In this misperception they were no different than the white majority of Chicago. After a three-year investigation of the 1919 riots, for example, the Chicago Race Commission concluded that the "primary beliefs held by whites" were, in order, that blacks are mentally inferior, immoral, predisposed to crime, physically unattractive, and highly emotional, given to "noisy religious expression." And white "secondary" beliefs, the report added, were that "Negroes" are "lazy . . . happy-go-lucky . . . boisterous . . . bumptious . . . overassertive . . . lacking in civic consciousness . . . addicted to carrying razors . . . fond of shooting craps . . . and flashy in dress."

In Harlem, the most popular nightspot for affluent whites was the Cotton Club, for dancing, drinking illegal liquor, and watching talented black performers in floor shows featuring beautiful black women. The customers included white gangsters, some of whom had supplied the bootleg liquor. But only the most influential African Americans were admitted. For the rest, there were plenty of other clubs—as well as illegal bars, gambling dens, and houses of prostitution.

In spite of Harlem's flourishing nightlife, the majority of the residents lived quiet lives, their leisure time spent at home or in church. These were the people to whom Bessie looked for support and if she visited the clubs of Harlem she was discreet about it. She used the New York office of the *Defender* as her headquarters because Abbott had ordered its manager, William White, to act as Bessie's agent. While there she was introduced to George W. Harris, editor of a black weekly, the *New York News*. Harris, a Harvard graduate, was a political power among New York's African Americans and alderman of the city's Twenty-first District.

One Sunday, with the alderman as her escort, Bessie went to the Metropolitan Baptist Church where she had been invited to speak by the pastor, the Rev. W. W. Brown, whose congregation numbered 2,500 members. Harris introduced Bessie to the assembly. In her first speaking engagement before a large crowd, Bessie displayed the eloquence she had learned as a child listening to Baptist preachers in Waxahachie. She described her first trip to France for flying lessons and said she was going back to take delivery of a new Nieuport plane she had ordered. During her three months in Europe, she would also purchase other airplanes for her school of aviation in the United States. And when she returned to the United States, she promised, she would give exhibition flights at Mineola, Long Island, and instructions at her flying school in New York. When she finished speaking, the congregation gave her a standing ovation.

Alderman Harris immediately offered to look after Bessie's interests when she returned from France. Some questioned Harris's honesty. Tammany Hall supporter Fred Moore, editor of another black weekly, the *New Age*, wrote that Harris was so dishonest he was "unfit to hold public office." Meanwhile, the *Washington Bee* carried a Negro Press Service report censuring both Harris and Moore for "washing their dirty clothes" in public. The story noted that while Harris's record as alderman was called into question his defenders accused Moore of shoddy behavior both socially and in business. The following year Harris would be unseated by the Board of Aldermen's Committee of Privilege and Elections. But, honest or not, Harris became Bessie's new manager, taking over from Abbott's New York manager Wright.

On February 28 Bessie's ship docked at Le Havre. There she told a reporter for the Baltimore weekly *Afro-American* that she was en route to the French capital where she would be trying out a Nieuport biplane especially built for her. There is no evidence that she ever owned the Nieuport, but she did take advanced flight instruction in it during the two months she was there.

During this second stay in France Bessie may have encountered the racist attitudes of white American tourists visiting Paris. The Americans' attitude became a much-discussed public issue. Bob Davis, a black theater manager, wrote from Paris, "White Americans must check the 'color line' at the three-mile limit. American capitalists and the petit bourgeois, touring France, have caused a furor." Davis detailed how American tourists resented black French colonials dancing with white women and having equal access to cafés, restaurants, hotels, and buses. After concerts, he wrote, when white Americans stood to sing the American national anthem, black members of the audience, some of them Americans, remained silent until they finished, then sang the "Marseillaise."

The controversy became so heated that white American expatriates soon sprang to the defense of the tourists in a concerted counterattack. One expatriate spokesman even warned that the American colony in Paris would take steps to bring deportation of "Negroes who infest the Montmartre section . . . continually insulting, assaulting and robbing tourists." Bessie must have bridled at such racist attacks but she said nothing about them in interviews, directing all her comments to her accomplishments and hopes for a future in aviation.

Despite the rise in racism, Paris's African American colony continued to thrive, embracing in 1925 Josephine Baker, who became the most famous black American expatriate and an idol of the French. With an aviation school to establish in the United States, Bessie herself would not consider living in France. She stayed for two months—probably the time needed to finish her advanced course—and on April 24 applied for and received a visa for Holland at the Dutch consulate in Paris.

She had decided to call on one of the world's most noted aircraft designers, Anthony H. G. Fokker. Fokker, who was Dutch, had been married to a German woman and was living in Germany at the beginning of World War I. He was either persuaded or coerced into designing planes for the German air force, aircraft that matched or excelled that of the Allies. At the war's end, the wily Fokker moved his entire factory to Holland so suddenly that the Allied victors had no time to confiscate it as enemy property.

Fokker not only received Bessie, she said, but he invited her to visit his plant and to fly some of his planes. He also took her to dinner and, she said, "after considerable persuasion" on her part, promised her he would come to America to build planes and set up an aviation school that would be open to men and women "regardless of race, color or creed."

Bessie left Holland only a few days after Fokker, who had gone to the United States to sell airplanes. Obviously determined to make the most of her limited time, she left Amsterdam and arrived in Germany on the same day, May 24, 1922, which was also the day her Dutch visa expired and her German visa, acquired earlier from the German consulate in Amsterdam, became valid.

Fokker may have given Bessie a letter of introduction but, with or without one, she soon made friends in the aviation circles of

Germany, where she stayed ten weeks. In Berlin Bessie was filmed by Pathé News, first standing by her plane and then flying over the defeated Kaiser's palace. She secured a copy of this newsreel (since lost) and later showed it to audiences during her lecture tours. She also brought home a picture of herself, wearing a leather coat and aviator's helmet, standing in front of an airplane beside a handsome pilot at Aldershof airfield near Berlin. Kurt Schnittke, technical assistant to World War I ace Gen. Ernst Udet, identified the pilot as Robert Thelen, the ninth pilot in Germany to receive a pilot's license, and the plane as an LFG-Roland with a 160-horsepower engine. Schnittke added that Thelen had been teaching Bessie how to fly the plane as part of an agreement by which LFG would deliver thirteen of the aircraft to her in the United States, presumably for use in her aviation school. If, indeed, there had been such a purchase agreement, Bessie never received the planes nor was she held to any contract.

Bessie also displayed to reporters a letter signed by a Captain Keller who she said was a German war ace and a member of the Deutsche Luftreederei, a postwar organization that controlled air traffic. The letter praised Bessie for showing "unusual skill" during fifty flights she had made over Berlin in a German plane equipped with two 220-horsepower motors.

Bessie left Berlin the first week of August and sailed for the United States from Amsterdam on the S.S. *Noordam*. When the ship docked at New York on August 13, a number of reporters were waiting to interview her, including two from white newspapers. Her answers to their questions were designed to create the image of a beautiful, brave black woman who loved flying and wanted to teach her people. Bessie hadn't the slightest doubt that she was that woman. All she had to do now was to convince the public. And that would not be easy because as an aviator she was a threat to whites who cherished their racial superiority, and as a *woman* pilot she threatened the ego of black males.

Bessie realized that to make a living at flying she would first have to dramatize herself, like Roscoe Turner, the great speed pilot who wore a lion-tamer's costume when he flew and took his pet lion, Gilmore, along in the second cockpit. An amused public paid to see this bizarre partnership, providing Turner with the money he needed for his plane, its fuel, and its maintenance.

Speaking to reporters, Bessie now began to draw upon everything at her command—her good looks, her sense of theater, and her eloquence—to put her own campaign of self-dramatization into high gear. Some of the things she said were contradictory or even close to pure fabrication. She recognized that the press was her path to the public. But if the reporters were going to create her public image, she was going to sculpt its details. Everything she told them was purposefully selected to enhance the image of a new, exciting, adventurous personality.

Bessie told the *New York Times* reporter, for example, that she had learned to fly after going to France with a Red Cross unit during the war. Brigaded to a French aviation group, she persuaded the officers to give her flight lessons. She also told him that she made a series of flights in a Dornier seaplane at the German base of Friedrichshafen, flights requiring "unusual aeronautical skill."

In Germany, she told the same reporter truthfully, she flew over the Kaiser's palace at Potsdam. And at Staaken, the former German military airfield, she said, she piloted, alone and without previous instruction, the largest airplane ever flown by a woman. There is no verification for this.

In another interview she said she was applauded in both Berlin and Munich for her "daredevil skills" and met a number of German celebrities. Baron Puttkamer of Berlin invited her to dinner, and she was presented to the Lord-Mayor of Baden-Baden. No doubt some of her accounts were true but the confirmation that might have been in German newspaper and magazine articles was lost when the Allies destroyed German archives in the course of the saturation bombing of World War II.

Certainly some of Bessie's claims were falsehoods. A week after she arrived in New York she told a *New Age* reporter that she was awaiting delivery of a dozen Fokker planes she had ordered for her aviation school. But the planes never appeared. In later interviews she added England, Belgium, and Switzerland to her list of countries visited although her passport bears no visa stamps from any of them. In every story she gave her age as 24 although she was by now 30.

The widespread news coverage Bessie now began receiving brought her increasing recognition. And it added distinction to her old friend, Robert Abbott. Convinced she would be useful in his aggressive campaign to increase circulation, he gave Bessie a desk in the New York office of the *Chicago Defender* at 2352 Seventh Avenue and told his staff there to arrange an air show for her. To build a large turnout for the show, Abbott's staff publicized it as being ostensibly in honor of the Fifteenth New York Infantry, veterans of the all-black 369th American Expeditionary Force of World War I. The show would star Bessie as the "world's greatest woman flyer," protégée of the "world's greatest weekly," the *Chicago Defender*. It was scheduled for August 27 at Glenn Curtiss Field in Garden City, Long Island, with a program that included parade-ground drills by doughboys of the Fifteenth, music by the Fifteenth's band, and eight other "sensational flights [by] American Aces" in addition to Bessie's performance. For spectators who wanted to take sightseeing tours after the show, giant passenger planes would be provided.

That Sunday morning when Bessie looked out the window of her room at the Hotel Pennsylvania, she saw rain. It rained all day. She hurried to the *Defender* office where she was assured the show would not be canceled, only postponed until the following Sunday.

During the week it was again publicized by the *Defender*, this time to take place on September 3 at 3:30 P.M., when the "wonderful little woman" would do "heart thrilling stunts." Bessie had been booked for an exhibition flight in Chicago on Labor Day but

Above: Commemorative buttons struck before the official record of Bessie's birthdate was uncovered. She was actually born in 1892, not 1893. (Courtesy of Marion Coleman.) *Below:* Bessie with an unidentified friend. (Courtesy of Arthur W. Freeman.)

Above: Bessie's sisters, Elois and Nilus, in the 1930s. (Courtesy of Arthur W. Freeman.) *Below:* Pilot Arthur W. Freeman, Bessie's nephew, whose decision to become an aviator was made as a young boy after watching his aunt give an exhibition in Chicago. (Courtesy of Arthur W. Freeman.)

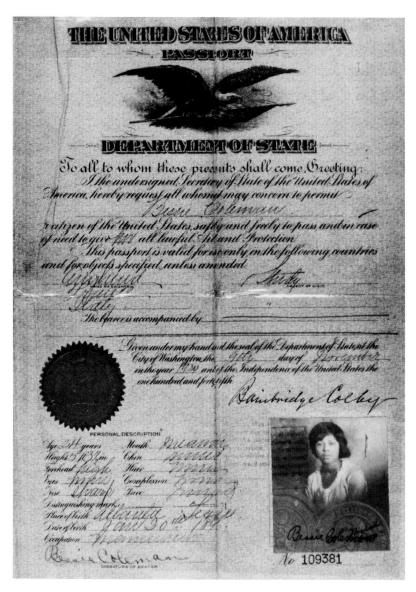

The front panel of Bessie's passport, issued in November 1920. (Courtesy of Thomas D. and Vera Jean Ramey.)

Above: Detail from passport. (Courtesy of Thomas D. and Vera Jean Ramey.) *Right:* Portrait of Bessie taken during her first stay in Paris in 1921. (Courtesy of Marion Coleman.)

Bessie standing in front of the Nieuport plane in which she learned to fly. The heavy flying suit protected her against the cold in the plane's open cockpit. (The Bettmann Archive.)

The picture used on Bessie's license issued by the Fédération Aéronautique Internationale. She was the first African American to receive one. (Courtesy of Arthur W. Freeman.)

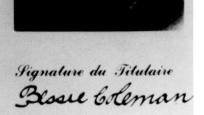

Bessie's flying license issued by the Fédération Aéronautique Internationale in Paris on June 15, 1921. (Courtesy of Marion Coleman.)

Above: Bessie with her pilot instructor, Robert Thelen, at Aldershof, near Berlin, where she took advanced aerobatic and flying lessons in 1921. (Courtesy of John Underwood.) *Below:* Bessie's mother, Susan Coleman, holding the silver cup given to her daughter in 1922 by the company of the Broadway musical *Shuffle Along*. (Courtesy of Arthur W. Freeman.)

G ernor Don̲a̲ y and May̲or̲ Thomas
Welcome Bessie Colen̲an to Columbus

GOVERNOR'S LETTER.
**STATE OF OHIO
EXECUTIVE DEPARTMENT
COLUMBUS**

Miss Bessie Coleman,   August 27, 1923.
    Chicago, Illinois.
Dear Miss Coleman:—

    The committee in charge of the Labor Day Celebration in the City of Columbus inform me that you are to be present and assist in the proper observance of that day.

    Therefore I extend to you a hearty welcome to the State of Ohio and to the City of Columbus and trust your stay may be both pleasant and profitable.

                Very truly your,
                VIC DONAHEY, Gov.

MAYOR'S LETTER.
**CITY OF COLUMBUS
OFFICE OF MAYOR**

Miss Bessie Coleman,   August 27, 1923.
    Chicago, Illinois.
My Dear Miss Coleman:—

    Word comes to me that you are to pay us a visit on next Monday to aid in the observance of Labor Day.

    Being familiar with your career and the skill, daring and courage you have exhibited on so many occasions, and knowing how your efforts have been recognized by the heads of many European governments, I deem it an honor and a privilege to welcome you to the City of Columbus.

    With the hope that your stay in our city will be pleasant and profitable, and again expressing my gratification that you are to visit us, I am,

Letters of welcome to Bessie from the governor of Ohio and the mayor of Columbus where she appeared in an air show in 1923. (Courtesy of Arthur W. Freeman.)

Above: The Curtiss JN-4 Bessie bought at Rockwell Field in Coronado, California, late in January 1923. Her ownership ended on February 4 when the aircraft stalled after takeoff. The plane was completely demolished and Bessie spent three months in a hospital. (Courtesy of Arthur W. Freeman.) *Left:* Bessie and an unidentified friend. (Courtesy of Arthur W. Freeman.)

Pictured with her Curtiss JN-4, Bessie wears the handsome leather coat and cap that she had made in France. (Courtesy of Arthur W. Freeman.)

ANOTHER BIG FLYING CIRCUS

Featuring

BESSIE COLEMAN'S

FAREWELL FLIGHT IN TEXAS

SUNDAY, JULY 12, 3:00 P. M.

HOUSTON AERIAL TRANSPORT FIELD

End of Main Street on Bellaire Boulevard

CAPT. MACKIE, famous dare-devil of the air, will do sensational stunts that seem impossible. Passenger planes will give you the thrill of your life and a bird's-eye view of the city. Ask any of the 75 who went up before.

O. P. DeWALT, Mgr.

Phones: Res.—P. 9163; Store Preston 7389.

Above: An advertisement for one of Bessie's aerial exhibitions in the African American weekly *Houston Informer* in July 1925. (Courtesy of State Historical Society of Wisconsin.) *Below:* Bessie in the tailored officer's uniform designed for her aerial exhibitions. (Courtesy of Fredia Delacoeur.)

Bessie's appearance in Savannah, Georgia, was the subject of a front-page story in the African American weekly *Savannah Tribune* when she arrived there in January 1926. (Courtesy of State Historical Society of Wisconsin.)

Left: Bessie in her show uniform wearing an officer's cap with wings. (Courtesy of Arthur W. Freeman.) *Below:* Extract from a letter by Bessie to African American film producer R. E. Norman. (Courtesy of the Lilly Library, Indiana University–Bloomington.)

W. Palm Beach Fla
Feb, 23, 1926

My Dear Mr. Norman
I just rec' your letter it was Not fowared to me as it should have been. Yes Mr. Norman I am More Than Sure my picture will go big in Colored houses I know this, as a Fact as my two News reels have drawn in house more so Than some Colored Drama

you may know what a real Film of 5 reels would mean you only have to ask the Mgr. at some of the Theatre in Fla. Tampa was not advertized "at all" But in St. Petersburg it was imposible to show have Have a chance to return. The picture that I want Filmed Maybe we could get together on it, yesterday, To day and tomarow It would be better if we joint put out the photo as I am not able to produce it independently now if you are exerested let me know also give me a price on directing 5 reels and eat shurg
Bessie Coleman 536 First St

The Elite Circle and Girls DeLuxe Club
expect you and your friends to enjoy
"An Aerial Frolic"
honoring
Miss Bessie Coleman
Sat May 1, 8:30 to 12 P. M.   Pythian Auditorium
Subscription 75¢
Music by the Imperial Jazz Orchestra

Invitation to the dance that was to be held in Bessie's honor on May 1, 1926, following her air exhibition in Jacksonville, Florida. She died in a crash the day before, while rehearsing for the show. (Courtesy of the Eartha White Collection, Thomas G. Carpenter Library, University of North Florida–Jacksonville.)

Above: Members of the Bessie Coleman Aviatrix Charity Club, formed in Chicago in 1928 to honor her memory, with Bessie's mother, Susan Coleman (*back row, center*). (Courtesy of Arthur W. Freeman.) *Below:* The author at the memorial stone at Bessie's grave in Lincoln Cemetery on Kedzie Avenue in Blue Island, Illinois (outside Chicago).

Above: Portrait of an aviator.
(Courtesy of Arthur W. Free-
man.) *Right:* Bessie Coleman.
(Courtesy of Bettman Archives.)

the paper rescheduled it in favor of the New York show "to please the thousands of persons of both races [who were] disappointed by the cancellation on August 27."

On September 3 Bessie made the "first public flight of a black woman in this country." The Fifteenth's men paraded and the band played jazz until Bessie walked out on the field. She wore the tailored uniform her nephew Arthur recalled so vividly years later, the Sam Browne belt, puttees, shiny boots, and leather coat, and she pushed up her goggles over her pilot's helmet so the crowd could see her face. The uniform was a combination of show business and Paris haute couture. The spectators loved it.

A writer from the Kansas City weekly, *The Call*, thought the crowd of from 1,000 to 3,000, depending on which reporter was counting, was "not as large as the occasion warranted." He expressed satisfaction, however, that it included important professional and business people, and added that the only disappointment was that Miss Coleman did not use her own plane. She flew a borrowed Curtiss JN-4 with a company pilot along with her on her first takeoff. Only after he had checked her out and passed her as satisfactory was she allowed to fly alone. This time she took off to loud applause and a rousing rendition of "The Star Spangled Banner." Though flying alone, she performed no stunts. Stunting was forbidden by Curtiss. Nor did any of the eight "American Aces" promised by the show's organizers appear.

Also on the program that day was Hubert Fauntleroy Julian, Bessie's equal in the art of self-dramatization. A flamboyant black aviator and parachutist from Trinidad, Julian made a 2,000-foot jump off the wing of a plane. Although he seemed to know no fear, he could not always be trusted to perform. At a show in Wilmington, Delaware, Julian disappointed Louis Purnell, a young boy who had come to see him play "The Flight of the Bumble Bee" on a saxophone while descending in a parachute, as advertised. Before the jump Julian rode in a limousine past the crowded grandstand. A few minutes later, and at a considerable distance, Purnell saw a figure leap from a plane and descend under an open chute, but he heard no saxophone. Purnell watched as the chutist landed and was picked up by a truck. The boy separated himself from the crowd and ran in the truck's direction. When he caught up with the vehicle pulling up to a stop behind an isolated hangar, he saw a man with a parachute in the truck bed, but the man was not Julian. Shortly afterward Julian reappeared in front of the grandstand, in the limousine, a parachute wrapped around him.

If competition from this tall (six feet, two inches), handsome man, who claimed to have been in the Canadian air force and prefixed his name with "Lieutenant," "Colonel" or "Doctor," annoyed Bessie, she gave no indication of it. The pilot of Julian's plane was Capt. Edison C. McVey, who frequently worked with the more colorful Julian. A year later McVey suffered a broken jaw and breaks of both legs and one arm in a flying accident but was back on the air-show circuit in less than twelve months. At the close of the show, McVey presented Bessie with a bouquet as the band played loudly and the audience cheered in approval. Julian, generally referred to in the press as "The Black Eagle of Harlem," had not detracted from Bessie's starring role.

New York newspaperman J. A. Jackson, a black columnist for the white-owned entertainment trade paper *Billboard*, wrote that Bessie was a conservative flier but a very good one. He quoted W. H. McMullen, assistant to the Curtiss Field chief pilot, as saying her "quick handling" of the Curtiss plane, an aircraft she had never flown before, "was very good." Jackson also noted that at Bessie's show a large passenger plane that took paying passengers on brief sightseeing flights over the area "did a big business all afternoon. Probably more people of color," he wrote, "went up that day than had ever flown since planes were invented."

Jackson was not only a very influential theater reviewer but was also busy organizing an association of black state fair promoters. Such fairs were gaining popularity in the rural areas of the South where entertainment for black audiences by black performers was severely limited by white authorities. On Jackson's strong recommendation, Bessie was engaged to appear at the Negro Tri-State Fair in Memphis, Tennessee.

In Memphis, Bessie's performance was billed as "the principal thrill" of the fair's opening day, October 12. Describing the occasion months later, Bessie said 20,000 spectators watched that first-day performance. (Twenty thousand was actually the attendance for the full three days of the fair.) The *Memphis Commercial Appeal* praised Bessie's performance as "nervy flying." And Jackson said Bessie proved "so great a draw that she should have been obtained for a dozen [shows]." Now, with two successful appearances behind her, Bessie Coleman, "the world's greatest woman flyer," was all fired up to return to Chicago and show her hometown what she could do.

CHAPTER 5

Pleasing the Crowds, Alienating the Critics

L ate in September of 1922 a triumphant Bessie Coleman, only three months shy of 31, came home to Chicago. In seven years she had engineered a personal transformation from penniless Southern immigrant to South Side celebrity. The former manicurist who as a young woman had observed and admired the prominent of her community along the Stroll was now herself one of the observed and admired.

For the first time her name appeared in the social notices of the *Defender* when a Sunday luncheon in her honor was given by a Miss Anthea Robinson of Forty-fifth Place and attended by a married couple and four bachelors. That same afternoon, September 30, as if to remind her of the fragile and ephemeral basis of her newly established reputation, a plane crashed in the middle of Main Street in Mount Vernon, Ohio, killing both its occupants.

"Queen Bess, Daredevil Aviatrix" was undeterred by the almost daily reports of fatal accidents in the high-risk occupation she had chosen. Two weeks after she came home she gave her first air show in Chicago, the one that had been originally set for Labor Day but which publisher Robert Abbott had to reschedule when rain delayed Bessie's first New York appearance.

In its pre-show publicity the *Defender* declared Bessie had "amazed continental Europe and been applauded in Paris, Berlin and Munich." For this show she would do four flights, starting at 3 P.M. at Checkerboard Airdrome, at Roosevelt Road and Fifth Street. More enthusiastic than accurate, the publicity could have been written by Bessie herself. One typical blurb ran:

> Her flight will be patterned after American, French, Spanish and German methods. The French Nungesser start will be made. The climb will be after the Spanish form of Berta Costa and the turn that of McMullen in the American Curtiss. She will straighten out in the manner of Eddie Rickenbacker and execute glides after the style of the German Richtofen. Landings of the Ralph C. Diggins type will be made.

The *Defender*'s "Berta Costa" was Bertrand B. Acosta, an American, not Spanish test pilot who would be one of Adm. Richard Byrd's crew to fly the Atlantic in 1927. The other Americans were Curtis McMullen, Ralph C. Diggins, and Edward Vernon Rickenbacker, all American aces of World War I. Charles Eugene J. M. Nungesser was a French ace and Baron Manfred von Richtofen was Germany's famous "Red Baron."

The same reporter wrote that on a second flight Bessie would cut a figure eight in honor of the Eighth Illinois Infantry and afterwards, presumably with Bessie at the controls, Jack Cope, veteran balloonist, wing-walker, and rope-ladder expert, would perform. On the fourth and last flight, Bessie's sister Georgia, younger by six years, was to do a "drop of death" as the parachutist. "No one has ever attempted this leap," the *Defender* trumpeted.

As far as Georgia was concerned it would be best if no one ever did. Certainly *she* would not. The night that Bessie came home with the costume she planned for Georgia to wear and blithely began to give her instructions on how to jump, Georgia shouted, "I will not, absolutely not, jump!"

"You'll do what I tell you!" Bessie shouted back.

"Who in the hell do you think you're talking to?" Georgia snapped.

They stood, toe to toe, glaring at each other until Georgia grinned and began chanting "Unh, unh, not *me!*" It was a chant the Coleman youngsters had used time and time again whenever Susan wanted them to do some distasteful chore. Bessie laughed, surrendered, and arranged for Cope to do the jump.

On October 7 and again a week later, the *Defender* ran a two-column advertisement stating that "The Race's only aviatrix" would make her initial flight at Checkerboard Airdrome on Sunday, October 15. The admission was one dollar for adults and twenty-five cents for children.

The *Defender* gave Bessie the publicity but the field and plane were provided by David L. Behncke, a white man. Behncke was a rare combination of shrewd businessman and aviation enthusiast who pushed his planes to maximum performance and encouraged other pilots to do the same. An Army Air Service instructor at 19, he was five years Bessie's junior but nevertheless owned Checkerboard, where he operated his own air express and charter service, fueling and repair station, and a sales office for new aircraft.

Because of this young entrepreneur, Sundays in Chicago had become a time for pilots to show the public their skills in racing, stunting, wing walking, and parachuting. Behncke, who would later become a commercial airlines captain, and president of the Air Line Pilots' Association for thirty years, had no reservations about Bessie's race or gender. He himself had replaced her in the Labor Day show at Checkerboard that conflicted with her New York appearance and won a speed derby by flying fifty-five miles in forty-five minutes.

On the Sunday of Bessie's Chicago debut the spectators, black and white, numbered about 2,000. Among them were her mother

Susan; sisters Georgia, Elois, and Nilus; nieces Marion, Eulah B., and Vera, and Nilus's son, eight-year-old Arthur Freeman. Arthur had always marveled at Bessie's stationery with its picture of an airplane on every sheet. Now, actually watching her perform, he was ecstatic. " 'My aunt's a flier!,' I thought, 'and she's just beautiful wearing that long leather coat over her uniform and the leather helmet with aviator goggles! That's my aunt! A real live aviator!' "

Bessie performed in one of Behncke's planes with considerably more dash and daring than at her first show in New York. She finished the first act in ten minutes before taking off again to do a figure eight in honor of the Eighth Infantry, turning and twisting the aircraft as if she had lost control, then soaring up again.

Sixteen months earlier Bessie may well have thought that nothing would ever surpass the joy of her first solo flight in France. If she later changed her mind and transferred pride of place to her Curtiss Field experience, that would hardly have been surprising. Her New York debut was charged with the double-barreled thrill of being her first chance to display her skills to her own countrymen and the first American performance ever by a black woman flier. Still, nothing brought as much elation to Bessie as her show at Checkerboard. Her solo flight had been for foreigners; her New York show for mostly strangers. Only Checkerboard could embody the unique satisfaction of performing in her hometown, flaunting her skills before her own people, her own friends, her family.

After the show passengers lined up for rides in one of five two-seater planes on the field. Bessie piloted one while Behncke and his assistants flew the others. They continued giving rides until dark.

Among the passengers that day was only one woman, Elizabeth Reynolds, who was so delighted with her brief flight that she asked Bessie for lessons. Applying with Reynolds was an unnamed male friend, a manufacturer of "extracts" and proprietor of a cigar store at East Forty-third Street. "Extracts" were often the ingredients used in the manufacture of alcoholic drinks forbidden by law. Most of the illegal liquor on the South Side was locally produced. But for wealthy clients who were willing to pay more for imported spirits, bootleggers were beginning to use airplanes as a means of eluding the law. Bessie could have made enough money from only a dozen smuggling runs to finance her flying school. But if she ever had any such offers she turned them down.

Only three weeks after her triumphant homecoming show at Checkerboard Airdrome, Bessie put her entire career as an aviator in jeopardy. Through the assistance of *Billboard* critic-columnist J. A. Jackson, who had given her rave reviews for both her New York and Memphis air shows, she had signed a contract to star in a full-length feature film. In a long *Billboard* article Jackson wrote that the projected venture was to be an eight-reel movie financed by the African American Seminole Film Producing Company and was tentatively titled *Shadow and Sunshine*. Bessie, he said, would be supported by twelve experienced actors, including Leon Williams, "one of the few race members of the movie branch of [Actor's] Equity."

One hundred extras would be hired for the film, which was to be jointly directed by script-writer Jesse Shipp and director Leigh Whipper. A Chicago businessman, Trueman Bell, was financial manager. Shooting was scheduled to start on October 18 in the hope of a Christmas release date.

But in another *Billboard* story scarcely three weeks later an angry Jackson wrote that Bessie "threw it up and quit cold." She did so after being told she would have to appear in the first scene dressed in tattered clothing and with a walking stick and a pack on her back, to portray an ignorant girl just arriving in New York. "No Uncle Tom stuff for me!" was her parting shot.

Jackson returned Bessie's blast with a caustic stab at Bessie's own humble beginnings in Waxahachie. "Miss Coleman," he wrote, "is originally from Texas and some of her southern dialect and mannerisms still cling to her." He followed this story with another in which he said the woman who would replace Bessie as the lead was an experienced actress with "unmistakable culture and social status which will be an asset to the company."

In an interview with Jackson, Peter Jones, president of the Seminole Company, said *Shadow and Sunshine*, originally intended to feature Miss Coleman, was delayed in production "because of the temperament of that young lady, who, after coming to New York at the expense of the company, changed her mind and abruptly left New York without notice to the director."

"Six autos filled with a cast of thirty people, two photographers and the directors waited in vain for two hours for the lady," Jackson wrote, "after which time Mr. Jones called upon her and was advised that she was too ill to accompany him to Curtis[s] Field for the few hours of outdoor stuff that was scheduled. That day she departed for Baltimore."

More than a broken film contract, one that he himself had had a hand in arranging, was behind Jackson's bitter denunciations of Bessie. In addition to his role as journalist and critic, Jackson aspired to be an impresario. At a meeting of the National Negro Business League he and Dr. J. H. Love, manager of the Colored State Fair of Raleigh (North Carolina), had founded the National Association of Colored Fairs. Both men were keenly aware of the desperate need for coordination in booking entertainers for the black state fairs that were just beginning to burgeon in popularity and they hoped their new alliance would end the existing state of booking chaos and confusion. But in that endeavor Bessie was to be of no help. Even before breaking her contract with the Seminole Company, Bessie had, in fact, managed to alienate both Love and Jackson as well as organizers of African American fairs in Virginia and North Carolina.

While still in Germany Bessie, acting on her own or through some unknown agent, communicated with officials of the Norfolk Colored Agricultural and Industrial Fair Association and agreed to appear at their Virginia fair on September 16. Bessie at the time was completely unaware of the existence of either Jackson or Love and could have had no way of knowing that her Norfolk booking had been arranged through their newly formed National Association of Colored Fairs. As early as July the black weekly *Norfolk Journal and Guide* enthusiastically reported, "Negotiations are underway to secure the wonderful colored aviatrix." And as late as September 8 the same newspaper featured a photograph of Bessie taken at her Curtiss Field exhibition in New York along with a statement that "Miss Coleman may appear at the Norfolk Colored Fair next week." But Bessie did not appear.

Since he had been instrumental in publicizing Bessie in *Billboard* and in setting up the Norfolk show, Jackson took Bessie's failure to appear as a personal affront. And now his new business partner, Love, was beginning to be concerned. For, when the Norfolk appearance was being negotiated, Bessie also assured Jackson she would play a date in Love's particular bailiwick of Raleigh, North Carolina. She promised she would send her final terms directly to Love but she never did get in touch with him. And, in the end, Bessie disappointed Love just as she did Jackson, flying instead at the twelfth annual Negro Tri-State Fair show on October 12 in Memphis, Tennessee.

Jackson now was publicly referring to Bessie as "eccentric," a common show-business euphemism for "unreliable." She had already had three managers in five months, he wrote. The first was William White, New York manager of the *Chicago Defender*. The second was Alderman Harris, owner of the black weekly *New York News*. And the third was "a white man whom she brought into the *Billboard* office. The lady," Jackson fumed, "seems to want to capitalize her publicity without being willing to work."

By this time Bessie had managed to offend men who already were or soon would be among the most powerful in the black entertainment world. Within a year entertainment entrepreneur and movie producer Peter Jones would be the manager of air-show fliers Edison C. McVey and Hubert Fauntleroy Julian and, in fact, owner of the airplane they used. Not only would Jackson continue his influential *Billboard* column until 1925, but he was also on the East Coast staff of the Chicago-based Associated Negro Press, a wire service used by most of the African American weeklies in the country. And Doctor Love continued his association with Jackson as promoters-bookers-organizers of black fairs throughout the United States.

When Bessie chose to do battle with Jackson, Love, and Jones, she was fighting on two fronts, in defense of equal rights for blacks in a white-dominated society and equal rights for black women in a male-dominated black society. Clearly her walking off the movie set was a statement of principle. Opportunist though she was about her career, she was never an opportunist about race. She had no intention of perpetuating the derogatory image most whites had of most blacks, an image that had already been confirmed by the Chicago Race Commission's report on the 1919 riots.

Bessie had refused to reinforce these white stereotype misconceptions or to further reduce the self-esteem of her own people by acting out on screen the role of an ignorant Southern black woman. As she herself had put it so bluntly when she stormed off the *Shadow and Sunshine* set, "No Uncle Tom stuff for me." Bessie was determined to bolster black pride. She was supported in this by a black press that seized on her image as both role model and grist for editorials promoting black equality.

The *Norfolk Journal and Guide* chastised the *New York Evening Journal* for stating in a long article that Bessie was from Europe and "took to flying naturally without any teaching." Bessie, the Norfolk editorial pointed out, not only was an American but was forced to take flying lessons in France because no one would teach her in her own country. The Norfolk paper further took the *Evening Journal* to task for stating that "the colored race should supply many excellent flyers" because blacks have a natural physical balance superior to that of whites and that blacks "usually ride a bicycle the first time they try." This latter claim, the Norfolk paper said in angry rebuttal, is a typical "Arthur Brisbane thought which runs through the editorial columns of all the Hearst newspapers . . . If we had more 'balance' along some other very important lines," the *Journal and Guide* concluded, "we would be hitting a greater stride in the race of races."

As to the balance Bessie sought—equal opportunity both for her race and her gender—the latter was far harder to attain. The majority of the nation's men, black or white, regarded women as the "weaker sex." Men headed households, governments, and businesses. Women were not even allowed to vote until the passage of the Nineteenth Amendment in August 1920. And women who aspired to careers were often mocked and even feared by men who saw them as a threat to their hitherto unshakeable faith in male superiority. The *Norfolk Journal and Guide*, for example, the same paper that had so readily praised Bessie as a black, was notably less ready to praise her as a black woman. In a miscellania column titled "Stray Thoughts" it included the following:

> **Blivens:** I see Miss Headstrong is taking aviation lessons.
>
> **Givens:** That so? Always had an idea that girl was flighty.

Bessie was ahead of her time as an aviator and as an advocate of equal rights for African American women. It would be three years before the rising tide of protest from black men was reflected by a blistering editorial in the *Amsterdam News* headed "Colored Women Venturing Too Far From Children, Kitchen, Clothes and Church."

The popular black weekly noted that "the biological function of the female is to bear and rear children," and said Kaiser Wilhelm of Germany had the right idea with his slogan of *"Kinder, Küche, Kleider, Kirche."* Calling black Washington, D.C., "a city of bachelors and old maids," the editorial focused on the fifty-five female teachers of Howard University, pointing out that whereas they had been raised in families totaling 363 children, or 6.5 children per family, the teachers themselves had only 37 children, or 0.9 per family. Worse still, it said, only 22 of the 55 were married, of whom four had one child and four had none. That the average age of the unmarried teachers was over 32 disqualified them as "good breeders," the editorial said, denouncing as "race suicide" what was happening in the nation's capital. "Liberalization of women," the editorial concluded, "must always be kept within the boundary fixed by nature."

Like the Howard teachers Bessie had crossed that boundary. She married late and was not living with her husband. She had no children and had become a pilot at a time when black male fliers could be counted on the fingers of one hand. She had a mind of her own. She was neither apologetic nor ashamed of being a so-called threat to the race or an affront to its men.

After breaking her film contract, Bessie left New York for Baltimore, where she gave an "interesting outline of her work" at the monthly meeting of the "Link of Twelve" at Trinity A.M.E. Church. She also went to Logan Field on the outskirts of that city to look over a number of airplanes but was reported to have found "none to her liking in which to take a spin."

Bessie's movie career was over before it began and her future as a stunt flier seemed in jeopardy as well. A good agent might have persuaded the Seminole Company's writer and director to reshape their script into a story stressing black pride and prowess, characteristics that would have had Bessie's approval. A good agent certainly could have eliminated her confusion over bookings for air shows. But none of the men Jackson named were really agents. White was a newspaperman put in charge of Bessie by Robert Abbott. Harris was a politician with a newspaper he used to promote his own interests. The unidentified white man is nowhere on record and may have been invented by Jackson.

Bessie returned to Chicago with little to sustain her beyond pride and obstinacy. There she launched into a search for new backers. If show-business people on the East Coast would not give her a break, she would look elsewhere.

CHAPTER 6

Forced Landing

In December of 1922 Bessie walked off that New York movie set without a contract, a sponsor, or a plane. Anyone else might have given up. She could not. Instead, at her next stop in Baltimore, she told a reporter that she was opening a school for aviators at 628 Indiana Avenue in Chicago.

From the moment she received her pilot's license in the summer of 1921 she had repeatedly talked of teaching the members of her race to fly. The school was not a publicity stunt, it was an obsession. An office at 628 Indiana was a start. The next step was to renew her contacts on the Stroll and at Checkerboard Airdrome, where David Behncke, Checkerboard's owner, would lend her a plane. But teaching in a downtown office was a far cry from founding a school. In addition to that office, she needed at least one plane of her own and a hangar for her aircraft and its maintenance, and she needed capital.

A giant step toward getting some came through one of her students, Robert Paul Sachs. Sachs was an African American, the Midwestern advertising manager for a California firm, the Coast

63

Tire and Rubber Company of Oakland. Soon after Sachs began taking lessons, Bessie proposed that for a fee she would go to California to "drop literature from the clouds" advertising the superiority of Coast Tires. The money would finance the purchase of a plane on the West Coast.

Bessie left freezing, windswept Chicago in late January on a train bound for sunny California. That state had already become a Mecca for the country's growing number of young fliers. In their fragile aircraft of wood and fabric and without radio or radar, they found Southern California's climate a godsend. For stunt fliers such as Bessie, California provided a built-in audience of unabashed admirers for the stars of the nation's newest pioneering industries—aviation and motion pictures.

A number of women were already benefiting from the recent trend of using women and airplanes in advertising, a fad that brought columns of free newspaper advertising. Two wing-walker/parachutists, Gladys Ingle and Gladys Roy, had found sponsors in real estate developers. Their free air shows over property up for sale drew crowds of potential customers. Six months before Bessie arrived in California, two young women pilots, Amelia Earhart and André Peyre, were flying for publicity and profit. In a printed testimonial, 22-year-old Earhart endorsed the airplane she flew for its builder, Bert Kinner. Peyre, a French actress, appeared with Earhart in exhibition flights to boost her own career in Hollywood.

Bessie knew none of these women. In the 1920s whites and blacks met socially only on the rarest of occasions. Whites generally ignored blacks but often held them up to ridicule and contempt, as in an article that appeared in the *Los Angeles Times* a month after Bessie arrived in that city. "Using his razor for social purposes cost Frank P. Hodge, a negro, $100 yesterday," the article said. Explaining that Hodge was courting a woman who objected to one of his remarks, the writer continued, "Hodge pulled out his razor and flourished it in such a manner as to abrade the lady's epidermis."

Yet in the face of such racism and without help from fellow woman aviators, Bessie could still match any competitor in skillful press agentry. She truly belonged in this land of flamboyant aviators and film stars.

For her first stop Bessie went straight to Oakland, to the Coast Tire and Rubber Company, to watch tires being made. In an interview after this inspection she "pronounced it the best, most modern and scientifically equipped plant ever seen in all her travels," although it is unlikely that she had ever seen a tire factory before. She declared herself so impressed she again volunteered, as she had to Sachs in Chicago, to "distribute advertising material from the clouds."

If the offer by this time was something less than spontaneous, the reporter nevertheless took Bessie at her word and wrote that, with every fifth person in California a car owner, "this method of advertising should be quite effective, stimulating the popularity of Coast tires and tubes." He also wrote that Bessie intended to open an aviation school in Oakland and would be giving exhibition flights in California and throughout the Pacific states.

Having paid a courtesy call on her new Oakland employer, and leaving a flurry of publicity for Coast Tires in her wake, Bessie headed south for Los Angeles to buy an airplane, most likely with Coast Tire Company money. The military kept a stockpile of surplus planes at the Rockwell Army Intermediate Depot on North Island at Coronado. By the time she arrived, there were only fifty left, many still in their original crates and not yet fully assembled. They were going for $400 each—or $300 each if you were rich enough to buy ten. Bessie settled for one—an early, almost obsolete Curtiss JN-4.

While at Coronado she was interviewed by a reporter for the *Air Service News Letter*, an influential professional publication that was widely read, especially by pilots. Acknowledging Bessie as "probably the only colored woman in the world who can pilot an airplane," he seemed more interested in her appearance than her accomplishments.

> Miss Coleman is a neat-appearing young woman who has discarded the shirt waist and short skirt for O.D. [olive-drab] breeches, leather leggings, Sam Browne belt and coat cut on the lines of Canadian officers. She says she went to France for two purposes, to drink wine and learn to fly. It goes without saying that she has been successful in flying, but we don't know yet her capacity or ability to drink wine.

Bessie told the *Newsletter* reporter that she had bought three planes, which she would arrange to have flown to San Francisco, where she would supervise their assembly and testing. Like many of his colleagues before (and after), he took her words for gospel and quoted her accordingly.

When Bessie spoke as a pilot she became an actress on stage, uttering fictional lines with total conviction. In her personal life she did not lie to friends and family, either telling the truth or remaining silent. At first she had simply embellished her stories in her search for recognition as a pilot. With that obtained, she now tended to even greater overstatement, saying whatever she thought might help her achieve her goal of founding an aviation school for her fellow African Americans.

She also called at the office of the African American weekly, the *California Eagle*, which proved as cooperative and obliging as the *Newsletter*. Taking eight years off her life, the paper described Bessie as an "enthusiastic, charming girl, only 23 years old, apparently unspoiled by the honors, social and professional, that have been showered upon her in the capitals of Europe."

Bessie's lying about her age was hardly unusual. Americans at the time considered a woman of 40 to be middle-aged. If any woman, especially a celebrity, could deceive the eye of the beholder, she might spend two decades claiming she was in her twenties. And Bessie, with her smooth face unblemished by any sign of aging, could deceive the eye of any beholder.

Never less than zealous in her persistent pursuit of the press, Bessie had prepared and always carried with her a press handout summarizing her accomplishments. Obviously she gave a copy to the *Eagle* correspondent. According to his story, Bessie had now flown in six European countries (up from three) and held German as well as French flying credentials. She again displayed the purported testimonial letter signed by the German Keller and told of being the guest of Anthony Fokker in Holland and of Baron Puttkamer and the Lord Mayor of Baden-Baden in Germany. She also claimed to be the only woman to hold an international flying license recognized "all over the world." Bessie was, in fact, the only American woman to receive her license directly from the Fédération Aéronautique Internationale in Paris. But there were other women, both American and foreign, who held licenses issued by aero clubs affiliated with the FAI.

In yet another interview, this one with the black weekly *Dallas Express*, she said she had come to California intending to open an aeronautical school in Oakland but the good people of Los Angeles had persuaded her to remain there instead. Their warm welcome included a dinner given by Mrs. S. E. Bramlett, of 1409 East Eighteenth Street, on February 1. All the guests were women, whom Bessie had begun to recognize as the dominant force for change in the black community and therefore the people most important to her plans.

The only hint of criticism in the otherwise laudatory press coverage of Bessie's visit up to this point appeared in an unattributed article in the Baltimore *Afro-American*. The story noted that "while Miss Coleman is an expert at the wheel of an airplane, she has made no long distance flights, confining herself mainly to exhibitions." This, it said, was the reason "the public is watching her effort" to break the flying record from Los Angeles to San Francisco. At the time, pilots were expected to attempt record-breaking flights, but Bessie had never said anything about wanting to break this or any other record. Quite possibly, "the public" was simply a euphemistic reference to the writer himself, who could well have been J. A. Jackson, the *Billboard* critic Bessie had so enraged by walking out on the Seminole movie contract in New York. Whoever wrote it, the implied criticism was certainly vitiated by the author's next statement that the distance between the two cities was about 80 miles and could be covered in an hour. A casual reader in Baltimore or almost any other part of the country might have overlooked that, but a Southern California reader, knowing that the distance was closer to 425 miles, would certainly have been brought up sharply by the glaring error. Breaking records was not on Bessie's agenda; opening a flying school for members of her race was.

As soon as her newly purchased plane was ready Bessie tried to arrange an exhibition flight at Los Angeles's Rogers Field, where Amelia Earhart had taken her very first airplane ride (as a passenger) just two years before. This plan never materialized, but Bessie found backers for another exhibition, on February 4, to celebrate the opening of a new fairgrounds at Palomar Park near Slauson Avenue.

As she left her room that Sunday at the Young Women's Christian Association's Twelfth Street Center for Colored Girls, she must have thought the worst of her struggles—learning to fly, soloing, getting her license, working for three different managers, and the ever-present pursuit of financial backing—was over. Now, finally, she had a plane of her own. And she had the use of a park for a show where she wouldn't just be one small part of a larger flying circus but the sole attraction. This time 10,000 people had gathered to see her, Bessie Coleman, just to "see a woman handle a plane."

But they never did see her. Moments after she took off from

Santa Monica for the short twenty-five-mile flight to the fair-grounds her motor stalled at 300 feet. Her newly purchased Jenny nose-dived, smashing into the ground. Airdrome workers ran to the demolished machine to pull Bessie's apparently lifeless body from the wreckage. She was alive but unconscious. A doctor who gave her emergency care at the crash site said she had a broken leg, fractured ribs, multiple cuts around her eyes and chin, and possible internal injuries.

On regaining consciousness, Bessie begged the doctor to "patch her up," just enough to enable her to fly the twenty-five miles to Palomar Park so that her waiting fans would not be disappointed. He refused and called an ambulance. En route to St. Catherine's Hospital on Pacific Avenue in Santa Monica, Bessie sent a message to her field manager, a Mrs. Bass, saying she would be along later—as soon as she could be "patched up."

Much more than a patch job was required. Bessie's broken leg had to be set in a cast from ankle to hip. She had three broken ribs, which made it painful for her to breathe. Many of her multiple cuts required stitches and caused massive swelling, especially around her eyes and chin. But far worse to Bessie than the pain and disfigurement was the loss of her one and only plane and of the chance to parade her skills before thousands of fans and ticket-buyers.

The accident itself was not unusual. In the preceding sixteen days one Los Angeles newspaper alone reported three similar crashes caused by motor failure right after takeoff, killing a total of five pilots. And two years later the Army Air Service's most distinguished aviator, Col. William F. "Billy" Mitchell, would crash just as Bessie had when his motor went dead at eighty feet.

What was, perhaps, at least somewhat unusual was the reaction of the crowd that had paid to see her perform and was waiting to see her at Palomar Park. Instead of the sympathy that might have been expected, many of the fans reacted in outrage, making strident demands for refunds "then and there." Bessie and her managers were accused of promoting a "bunco game." The charges were so widespread and virulent that one journalist, Dora Mitchell, a reporter for the *California Eagle*, wrote that it was "with shame for our own people" that there was a "most deplorable dearth of chivalry among our men and an utter lack of womenly feeling and sympathy among those of Miss Coleman's own sex." The story went on:

> a brave little Race girl was condemned without a hearing while she lay on a bed of pain, unable even to send a message . . . although such a message would doubtless have been received with sneers and incredulity . . . Certain people on Sunday night even declared this poor girl's injuries to be a punishment from on high for the sin of attempting to fly on Sunday.

Battered and bandaged but unwilling to admit defeat, Bessie sent this telegram from her hospital bed to friends and well-wishers:

> TELL THEM ALL THAT AS SOON AS I CAN WALK I'M GOING TO FLY! AND MY FAITH IN AVIATION AND THE USEFUL OF IT WILL SERVE IN FULFILLING THE DESTINY OF MY PEOPLE ISN'T SHAKEN AT ALL.

What Bessie lacked in grammar she made up for in ambition. And she insisted she would carry on her plans for a school where "our boys may acquire the mastery of the air."

A week later, a reporter who described her as "undaunted by a broken leg" recalled that in a lecture before the accident she had said, "I am anxious to teach some of you to fly, for accidents may happen. I may drift out and there would be someone to take my place." This was not all self-dramatization for the sake of publicity. Like all circus and stunt fliers Bessie knew she was engaged in a high-risk occupation. And she accepted those risks because she loved flying and regarded her participation in it as a way of generating self-respect for her race.

When news of the accident reached the family, Elois said, "Well, that's the end of it." Susan, who once claimed she "had thirteen children, raised up nine and one of them was crazy," knew her Bessie. "Oh, no," she told Elois. "That's only the beginning."

Bessie's leg took far longer to heal than she had anticipated, but she got some badly needed support from her former Chicago student and Oakland business sponsor Robert Sachs. Sachs wrote a long letter on her behalf, which was published in the *Eagle*. He, too, parroted all the background information Bessie had given him, including her age as 23, and added:

> Her race should be proud of her accomplishments, as she is the only colored woman pilot in the world—and there are no colored men flyers at all. And with all her hard work and her great record of accomplishments, she has always borne her race in mind and worked for its advantage. One of her plans for the future is the establishment of a school of flying especially for colored people.

Her work, Sachs continued, had been delayed "as much by lack of funds as by her personal injuries" and "worry over financial problems" was likely to delay her recovery.

A subscription had been taken up at the company's office, he said, and he was offering the African American people of California "the opportunity to really come to the assistance of the race." Sachs said black churches and other organizations might raise funds from their members, and those who wanted to contribute to Bessie directly could send money to her at the hospital, to a Mrs. Melba Stafford at 939 Willow Street in Oakland, or to the office of the *Eagle*.

Bessie was down, but far from out. Without capital or a single airplane, she placed a paid notice in the *Eagle*, a contract for the Coleman School of Aeronautics. Its terms included a student tuition fee of $400 in advance, $25 on signing the contract, and the remainder to be individually agreed upon in monthly installments. In return she promised to provide competent instructors and to put all the money paid in escrow until the lessons started. The student would be responsible for all injuries to his or her person, and Bessie would be the judge of his or her ability to operate an aircraft. She would give the student every available means of instruction and equipment to demonstrate that ability.

Her advertisement of the contract came to the attention of the *New Age,* a black newspaper clear across the country in New York, that published an editorial headlined "Queen Bess Opens School." The rules of the contract, the editorial said, could be interpreted as meaning that "broken bones and other dangers are at the risk and expense of the flyer in embryo. There is nothing like having such details understood in advance."

Bessie's contract was a conventional one for the times. In fact it was virtually identical to the "Conditions de l'Apprentissage," the contract given student pilots at the famous Caudron School of Aeronautics at Le Crotoy. Bessie may have hoped that the contributions received from the collections by Sachs, together with advance payments from a few students, would be enough for a start. It was not.

It was early May, nearly three months after the accident, before Bessie was able to leave the hospital, her leg still in a cast. "Although hardly able to put the weight of her body on the injured member," said one newspaper account, "the gritty and progressive woman has, during the past week, completed a motion picture started on the day before her unfortunate fall." Sachs too had mentioned Bessie's making a motion picture in Los Angeles but neither report went into further detail and there is no confirmation that a movie was ever released, made or, indeed, even begun.

For the remainder of May Bessie stayed at the Los Angeles home of a Mrs. S. E. Jones on West Fifty-seventh Street. Still striving to raise capital for her school, she arranged to give five nightly lectures between May 5 and 12 at the Ninth Street branch of the

Young Men's Christian Association. At these she showed films of her flights in Europe and the United States. Wendell Gladden of the *Eagle* wrote that one of the reels, from Pathé News, had already been shown at a popular theater in Los Angeles. It showed Bessie standing beside her plane in Germany and later airborne, "soaring high above the Kaiser's palace." The admission charge for her combined lecture/picture show was twenty-five cents for children and thirty-five or fifty cents for adults. Bessie, of course, did not get all the money. Some of it went to the YMCA.

Bessie left Los Angeles for Chicago in June without a plane, without a job, and with very little money. Once again she would just have to start over.

CHAPTER 7

Grounded

When Bessie Coleman returned to Chicago in late June 1923 she was determined to halt the downward course of a career that had lost its momentum during the months she had spent recuperating from the crash in Santa Monica. It was not going to be easy. The *Chicago Defender* appeared to have lost interest in her. After a brief notice about her return, the paper failed to publicize her first post-crash air show, scheduled for September 3, Labor Day, in Columbus, Ohio. This time a white daily, the *Columbus Evening Dispatch*, gave her the advance publicity she needed. The *Dispatch* billed her as "the only colored aviatrix in the world," who would stage a number of flying stunts at Driving Park, a racetrack, on Labor Day.

Whether owing to an unusually credulous reporter or a copy-short editor, the *Dispatch* story used almost all of Bessie's prepared publicity handout, claiming she had given exhibition flights in Europe and that she was raised on a ranch in Texas where she was "born twenty-three years ago." The *Dispatch* reported that the show would include airplane, automobile, and motorcycle races and stunts by a man billed as "Daredevil Erwin" and by two women, Iona McCarthy and Bessie Coleman. Erwin's repertoire included hanging by his teeth from a strap suspended beneath an airplane. McCarthy, who did triple parachute leaps, said she would try to land on the wing of another plane. The pre-show publicity did not say what Bessie would do. Presumably just being black, female, and the first of her kind to fly a plane was thought to be sufficiently intriguing.

But the Labor Day show that had promised to be a bonanza for Bessie's flagging career was rained out. She sat all day at Driving Park, along with her colleagues, waiting for the weather to clear. While she waited, more than 2,000 people—almost as many women as men—gathered at the state fairgrounds a few miles away for an all-day celebration of the Ku Klux Klan. Their enthusiasm undampened by the rain that was washing Bessie's show off the calendar, it was primarily an initiation festival, men being inducted into the Klan itself and women into the Ladies Auxiliary. Called "naturalization," the initiation preceded the marriage of two Klan members by a Klan minister. Music for the festivities was provided by an orchestra of Columbus Klan members. Almost everyone, including the musicians, dressed in the ghost-like white hoods and robes that inspired fear in so many African Americans.

Leaving rain-soaked Driving Park and the state park full of Klan members behind, Bessie returned to Chicago, where she called at the *Defender*'s office to announce that her Columbus air show had been rained out but her plans to open an aviation school in Chicago were "well underway." According to the *Defender*, Bessie also stated that she would give her first flight in Chicago "as soon as her machine arrives in the city" and the plane itself would be put on exhibit at the Eighth Regiment Armory. The paper affirmed that Chicago's favorite aviator had completely recovered from her accident in Los Angeles and said "rumors were afloat at the time that some mechanics tampered with the steering apparatus in an effort to keep her from gaining the recognition due her." The *Defender* was the only paper to publish such rumors.

On September 9 Bessie returned to Columbus for the air show that had been rained out a week earlier and performed before an enthusiastic crowd of 10,000. Back in Chicago once more she made another announcement to the *Defender,* saying she would give "a farewell flight" in the city before leaving for "a tour of the south." The show was advertised for Sunday, September 23, at the Chicago Air Park at 63rd Street and 48th Avenue. Yet the following week's *Defender* had no story about it. It may have been canceled because the plane she had offered to put on exhibit never arrived. Worse yet, there would be no Southern tour for Bessie, who had run head-on into a series of changing managers and broken contracts.

Bessie's contract troubles became public in the spring of 1924 when the *Afro-American* ran an unbylined story headlined "Aviatrix Loses Another Manager." In a style strongly resembling that of her enemy, J. A. Jackson, it described Bessie as "the Colored girl who has been presenting herself as an aviatrix for the last two seasons" and accused her of accumulating "a long list of incomplete contracts and an almost as lengthy list of managers and agents." Naming the managers and agents as D. Ireland Thomas, M. C. Washington, and, the most recent, Raymond Daley, the writer asserted, "we understand that Washington has not added to his prestige in Ohio by his experience with the temperamental aviatrix."

By that spring, Bessie had already had five known managers in twenty-eight months—*Defender* editor William White, New York alderman George W. Harris, and the three named in the *Afro-American*'s article. Her sister Elois added a sixth, David Behncke, to the list, although he could do little more for Bessie than book her in local air shows at Checkerboard, along with another black pilot, Edward Young, from Iowa.

Whether or not she was as "temperamental" as the *Afro-American* writer claimed, Bessie certainly was strong-willed and independent enough to insist on having a say in her bookings. Her African American agent/managers, on the other hand, were equally determined, raised in a business tradition where the manager led and the client followed, especially if the manager was male and the client female. That she and they should clash was probably inevitable.

Purposeful though she was, Bessie now found herself in Chicago at the end of a long series of misadventures with no employment in sight. The loss of her only plane in California, her long hospitalization, and the alienation of her managers—first on the East Coast and now in the Midwest—seemed to be having their cumulative effect. She appeared ready to give up, telling her sister Elois that she had decided to take "a good long rest." Bessie rented an apartment at Forty-second and South Parkway (now Dr. Martin Luther King Jr. Drive) and, according to Elois, "took her furniture out of storage, and clad in coveralls, went about cleaning the place up."

Despite her being a celebrity, there was no niche in black Chicago into which Bessie could fit and feel really at home. She was not a businesswoman like Madame C. J. Walker or an educator like Dr. Mary McLeod Bethune. She had none of the editorial skills of Ida Wells Barnett. Even as an entertainer she had nothing in common with the Chicago celebrities of the time—Ethel Waters, Bill "Bojangles" Robinson, Alberta Hunter, Bessie Smith, or Lil Hardin, who had just married Louis Armstrong.

Neither was Bessie interested in the activities of the Pilgrim Baptist Church, which were so essential to her mother's social life. However, she did accompany Susan there occasionally and once, in a lower-floor lecture room, was introduced to Marjorie Stewart Joyner, friend and employee of Madame Walker. An observant and gregarious businesswoman, journalist, and community activist, Dr. Joyner remembered Bessie as "middle-class, like me and most of the people there that night. She dressed nicely—nothing too showy—and was well-spoken. She was trying to *be* somebody. Not like people who say they want to be somebody but don't really try. We tried, Bessie and I, and we succeeded."

Dr. Joyner was right. Bessie was already a South Side celebrity. So was Joyner, who was active in the Cosmopolitan church, Bethune-Cookman College, the National Council of Negro Women, and the United Beauty Schools Owners and Teachers Association, as well as in raising funds for any number of Chicago charities. (In 1990 Chicago's mayor Richard M. Daley, Jr., declared October 24 to be Dr. Marjorie Stewart Joyner Day in honor of her ninety-fourth birthday.)

"But we weren't upper class," Dr. Joyner added. "True society people were all doctors, professional people or very wealthy business men and their families. That upper crust didn't want to help Bessie. Robert Abbott helped because he was interested in news and Bessie was news."

There was no place for a manicurist-turned-pilot in Chicago's black society, not when some black clubs and churches would not accept members unless one could see "the blueness of their veins through the skin of their wrists." "The right shade of black is yellow" was the general rule. So Bessie turned to her friends and family.

At home, her experiences as pilot and world traveler had affirmed Bessie's pivotal role in the family. Her apartment was where all the Coleman women met. A center for both happy reunions and fierce arguments, it was open to all family members. In their eyes she was famous and because of that all the Colemans had gained status. Her Southern, black, rural heritage of allegiance to kin was expressed in the support and loyalty she gave all of them.

The wife of Bessie's brother, Isaiah, who had gone to Canada, received a warm welcome when she came to Chicago on a visit. Bessie was charming, she said, amusing her one evening by giving a series of impressions of Asians, Caucasians, and blacks, all done with the skillful use of theatrical makeup, wigs, and a wardrobe that included French frocks and a beautiful silk negligee. "Your Aunt Bessie is very pretty and very clever," she said. "Somehow she can even make herself up to look like different nationalities."

Bessie's apartment became a second home for her nieces and nephews—a home she herself had never had with hardworking Susan who had neither the time nor money to give her children anything except the bare necessities. Bessie hugged and kissed her sisters' children and let them play her records on the wind-up Victrola. A marvelous cook, she made them meals, ate with them, and taught them table manners without scolding. She bought them clothes and toys and gave them spending money. Bessie gave Elois's daughter Vera, who was eleven, something Bessie herself must have longed for as a child, a room of her own, in Bessie's apartment. "Back when I was little, when she was still doing manicures," Vera said, "she'd give me money for the movies and walk down to the corner of Thirty-fifth to watch when I crossed the street. Then, when she came back from California, I got to live with her."

Vera's brother Dean recalled that "Vera had a little dog, a white Spitz and she loved that little dog so much she named it Bessie."

Their favorite aunt also paid the tuition for Georgia's daughter, Marion, to attend kindergarten at the Little Shepherd School on a site now occupied by a public school at Forty-ninth Street and Indiana Avenue. And it was Aunt Bessie who rescued Marion from her grandmother's insistence that the child remain at the Pilgrim Baptist Church every Sunday for the entire day. "I was so restless and bored," Marion said, "until Bessie got me excused for the afternoons. She gave me money for the movies and two dimes to ride the El to and from the Peerless Theater. I saw a movie almost every Sunday afternoon."

In contests of will Bessie prevailed over her mother and sisters—except for Georgia. Georgia was a fun-loving "flapper" of the Twenties, a devotee of the Stroll's nightlife. She coveted Bessie's elegant clothes and often took things from her closet without permission. Bessie's wardrobe was mostly conservative, but it did include a glittering, red, beaded designer creation from Paris. One night when Susan was at Bessie's apartment, she saw Georgia taking the red dress out of the closet. "Bessie will kill you!" Susan shouted. Georgia just grinned. She was in it and out and dancing at Dreamland before Bessie could catch her.

Bessie went out, too, to many of the same clubs on the Stroll, but her evenings were more sedate than impetuous Georgia's and her escorts were older and richer. One was Kojo Takalou House-mou, the handsome visiting prince of Dahomey, whose company was much sought after by African American society throughout the United States. Kojo had been invited by the National Association for the Advancement of Colored People to visit the United States under the direction of the association's secretary, Walter F. White. The invitation was the NAACP's way of showing its disapproval of Marcus Garvey's Universal Negro Improvement Association, the back-to-Africa movement that claimed thousands of ardent follow- ers. Kojo, who said he had a doctorate degree in philosophy from the University of Paris, headed an organization of his own in French Colonial Africa known as the League for Defense of the Black Race.

Bessie gave a tea in the prince's honor, charming him and his retinue by speaking French. Other of Kojo's evenings were far less pleasant. One night at a downtown Chicago restaurant, he was refused service by the waiters because he was black. When the prince objected, the restaurant summoned the police and, in the ensuing melee, the guest of honor lost two gold teeth.

Bessie had other men friends but aside from "Uncle Claude," Bessie's husband, who still called on her, Bessie's nieces did not know their names. One is thought to have been her old acquain- tance from her manicurist days, Jesse Binga. By this time Binga, real estate magnate, owned the lot at the corner of State and Thirty- fifth on which he was to build "the most modern bank and office building owned by colored people in the world."

As for Robert Abbott, the publisher seems to have distanced himself from his one-time protégée. Possibly he was concerned that Bessie's alienation of so many managers and agents would cut into the *Defender*'s advertising revenue, a large share of which came from African American theatrical notices. Between October 1923 and May 1925, his newspaper published only one item with Bessie's name in it. Datelined New York, May 12, 1924, it was an announcement by Maj. Lorillard Spender that Dutch plane designer Anthony H. G. Fokker had leased the former Witteman Aircraft factory in Hasbrough Heights, New Jersey, to build planes in the United States. Although Bessie was not at all integral to the story, inserting her name would most assuredly increase local interest in Chicago. The *Defender* identified Fokker as the man who had taught Bessie to fly and again quoted Bessie as saying that Fokker had promised, "after considerable persuasion on her part," to establish an aviation school for men and women, regardless of race, creed, or color, which would charge the least possible tuition. A year later Spender and Fokker formed an Atlantic Aircraft Corporation partnership but nothing ever came of the school.

Languishing in Chicago with no job and no capital, Bessie was not the only pilot whose luck had run out. Hubert Julian was placed on probation for six months for parachuting inside the city limits of New York. When the probation period was over he lectured to paying audiences such as one at the Attucks Theater in Norfolk, Virginia, in a struggle to raise money for a "world flight."

The following July Julian was ready—or thought he was. His takeoff was delayed for hours because the builders of his seaplane, the Chamberlin Roe Aircraft Corporation, demanded the $1,400 he still owed on the $8,000 airplane. Julian stood on top of a taxi pleading for donations from the crowd gathered to see him leave from the Harlem River at 125th Street. Assisting Julian were several friends who sold photographs of him and his airplane to onlookers until two admirers, a minister and an undertaker, pledged the final $500.

Meanwhile, the plane had become mired in mud as the tide receded. Mechanics dragged it into deeper water where three attempts to start the motor failed. On the fourth try it started but the plane hit the rocks, now exposed at low tide at 140th Street. From there the aircraft was hauled by boat down the Harlem River to the East River. Immediately after takeoff a pontoon dropped off and fell into the water. The plane tilted, then crashed into Flushing Bay, six miles from where it had started. The whole affair was a fiasco, but one that easily could have happened to any flier in that era of primitive planes and capital-starved pilots.

Just as brave, as daring, as determined, and perhaps as foolish as Julian or any other of these aviation pioneers, Bessie could not give up flying. She kept looking for a part in an air circus—anywhere. After eighteen months of searching she finally succeeded, lining up a series of lectures and exhibition flights in Texas.

CHAPTER 8

Texas Triumph

In May of 1925, twenty-eight months after her plane was shattered on a California airfield, and with it all of her plans, Bessie
arrived in Texas. Leaving behind her the bad press of the East
Coast and the bad luck of the West, she returned to her first
home—Texas. For her home base she chose Houston and for her
first performance a lecture at the Independent Order of Oddfellows
(I.O.O.F.) temple on May 9 with films of her flights, a format she
had developed during her lecture series at the YMCA in Los
Angeles.

Bessie had lost none of her ability to glean good advance press
notices, one of which, in the *Houston Post-Dispatch*, a white daily,
quoted the "widely-known daredevil pilot" as saying "the Negro
race is the only race without aviators and I want to interest the
Negro in flying and thus help the best way I'm equipped in to uplift
the colored race." Her great ambition, she went on, was "to make
Uncle Tom's cabin into a hangar by establishing a flying school."

She also said she was 23 instead of 33, a college graduate who
had received flight training in Amsterdam, Berlin, Paris, and "else-

where," and had recently returned from her latest trip to Europe where she had given exhibition flights.

Five weeks later Bessie made her first flight in Texas, on June 19 in Houston. Her Chicago sponsor, David Behncke, may have had a hand in securing the Texas engagements, but well-known local flier Capt. R. W. Mackie was in charge of the Houston show, which was promoted by O. P. DeWalt, owner of the Lincoln Theater, and J. M. Barr, secretary of the Negro Amateur Baseball League. Mackie himself probably provided the postwar Jenny powered by an OX-5 engine that Bessie flew. Because of the extensive advance publicity and because Bessie's appeal was so widespread that it cut across all prevailing color lines, DeWalt, Barr, and Mackie had to provide "special reservations" (that is, separate seating) for white patrons.

An extra drawing card for Bessie's first Texas air appearance was that her show was scheduled for the same day that African American Texans celebrated as the most important holiday of the year—Juneteenth, the anniversary of the day blacks in Texas achieved their freedom. Although President Lincoln had issued his Emancipation Proclamation on New Year's Day, 1863, it didn't take effect in Confederate-governed Texas until more than two and a half years later—June 19, 1865—when Union troops under Gen. Gordon Granger marched into that state from the North. Also, just a week before Bessie's Houston show, the U.S. House of Representatives, after a three-hour floor fight led by Representative Thomas Lindsay Blanton of Texas, defeated a proposal to make Lincoln's birthday a legal holiday. Thus Bessie's Houston appearance came as a thrilling defiance of such bigotry.

The air-circus heroine, born in a dirt-floored cabin in Atlanta and raised by a single parent in Waxahachie, was as much a celebrant as her audience. That Juneteenth the one-time manicurist did not *tell* her people what they could do if only they were determined enough, she showed them. Ascending high above the clouds, she stalled her motor and dove to within a few feet of the ground before pulling up as the crowd let out a roar of relief and admiration. Interspersed with dives were barrel rolls, figure eights, and loop-the-loops. Her aeronautic pyrotechnics sent a direct message to the spectators below: If I can do it, so can you!

The *Post-Dispatch* ran two advance stories on the show. The first told of the city's black ministers making flights over Houston along with supervisor Mackie. It further reported that Bessie was "said to be daring as well as skillful" and a huge crowd was expected to witness her flights. The second story, the day before the show, noted that Bessie was one of the big attractions and that "thousands of spectators of both races expect to be on hand to see stunt flying promised by aerial daredevils." Citing numerous clippings from other newspapers all over the country, it further lionized Bessie and propagated her message:

> This negro woman attracted attention all over the country for her efforts to interest the negro race in aviation. She is well educated and spent a number of years abroad. She says the negro race is the only one in the world that has not yet developed aviation and she is trying to get negroes flying.

The fact that she could wrest such laudatory and extensive coverage from a newspaper generally regarded as a spokesman for the white Texas establishment, and which steadfastly refused to capitalize the word "Negro," was more than just a major triumph for Bessie. It was sure evidence of the revolutionary change, however slow, subtle, or imperceptible, she was effecting on white attitudes toward blacks, at least in her home state of Texas. However, almost as if embarrassed by its coverage of a story that many of its readers probably considered of primary interest only to blacks, the next day the *Post-Dispatch* ignored Bessie's show completely and featured a "shooting and cutting rampage in the city" in its reports on Houston's Juneteenth black community celebrations.

Bessie's show was not as professional as those presented by entrepreneurs who could afford better equipment or by high-priced crowd-pleasers such as Mabel Cody, who climbed from a speeding boat up a ladder suspended from a plane overhead, and Gladys Roy, who danced the Charleston on the upper wing of an airborne biplane. Still, the combination of an opportunity both to see the world's only black woman flier and to ride in an airplane oneself proved an irresistible rousing success. Thousands came to the old speedway auto racetrack at the end of Main Street, renamed Houston Aerial Transport Field in honor of the occasion, to see Bessie perform.

Juneteenth 1925, Bessie's flying debut in Texas, was more than just a big day for Bessie and for Houston. It also marked an event of historic significance for her country. Although overlooked or unnoticed by virtually everyone else at the time, the city's leading black newspaper, the *Houston Informer*, recognized and reported that it was "the first time colored public of the South ha[d] been given the opportunity to fly." Calling it "the biggest thrill of the evening and of a lifetime for that matter," the *Informer* noted that "about 75 of our fearless citizens, most of whom were women," climbed aboard one of the five small passenger planes available to get "a birds-eye view of Houston from the sky." The paper later identified a Reverend Harrison of the Antioch Baptist Church as "the first colored person in the city to hop off for a ride. Rev. Woolfolk of Trinity Methodist Episcopal, not to be outdone, took little Miss Woolfolk with him." Not everyone at the show, however, was quite so courageous. "One of our leading ministers reneged on his wife when she requested him to escort her for a ride."

The Houston African Americans who dared that day did indeed need courage, not so much to fly as to make themselves conspicuous in a mixed crowd of Texans. Texas blacks at the time were leaving the state in record numbers to escape the worst Jim Crow laws in the country. A contributor to the Baltimore *Afro-American* wrote that although there were two trains a day going from Houston to San Antonio, black passengers were not allowed to ride the morning train, which had Pullman cars. They had to wait twelve hours for the night train and were confined to a partitioned half of the men's smoking car, which had only one toilet for both sexes. When the reporter went into the dining car a green curtain was thrown up around him to separate him from the rest of the passengers, and, at the end of his meal, he was handed not the regular dining check but a special one identifying him as a Negro.

The Baltimore writer's experience notwithstanding, Houston African Americans that day not only dared to fly but, according to the *Informer*, were left "clamering for another [chance to do so] soon." And, no doubt speaking for itself as well as for its source, the *Informer* quoted "the management" of Bessie's show as saying it "was more than glad to see [Houston African Americans] disprove the assertion that the Negro is afraid to fly." The paper concluded that the day had been an unmitigated success, "the only difficulty experienced" being how to accommodate in the few aircraft available everyone who wanted to go up and to fly them "as fast as they wanted to go."

Bessie's Houston debut was so successful, in fact, that she was immediately booked for three more appearances, one at Richmond, Texas, on July 10 at a meeting of the Southeast Central Baptist Association, a second on the following day at a picnic of the Southern Pacific railway's African American employees in Houston, and a third on July 12, also in Houston.

Promoter O. P. DeWalt predicted that thousands would attend the Houston show on July 12, and they did. The *Chicago Defender*'s reporter wrote that "she took off with a perfect start in the OX5 plane [a JN-4 with an OX-5 engine], circled the field several times, then 'gave it the gun' and spiraled and looped to the satisfaction of the roaring crowd." Mackie closed the exhibition with his own "daredevil act" of aerobatics and a parachute jump, but Bessie was the undisputed star of the show.

Although the *Houston Informer* reports of Bessie's Juneteenth show did not specifically say so, they strongly suggested (by their style and their occasional judicious use of the word "volunteer") that the sightseeing flights for Houston's leading black clergymen had been offered free as an inducement to the rest of the spectators. This time, however, with the management making an extra sixth passenger plane available, the crowds lined up to pay twenty-five or fifty cents apiece for their flights over the city.

In Houston, Bessie appealed to African American women to take an interest in aviation. As she had begun to do in California, Bessie would continue to focus her attention more on women than on men, convinced that black women would be more effective activists than men in furthering the participation of African Americans in the political and educational systems and in the struggle for equal rights. DeWalt agreed, claiming only women truly "appreciated the fact that it took a woman to put the Race on the map." In spite of her pride of race, Bessie was pragmatic in appealing to African American women rather than men. She knew that the white, male-dominated society found it easier to accept assertive black females than to accept the same characteristic in black males.

The next recorded show for Bessie was on August 9 in San Antonio, where she appeared again with Mackie at the old San Antonio Speedway. This time the parachutist was a local African American woman, Liza Dilworth, of 1417 Crockett Street. She would be the first black woman to make a jump, Bessie said. Later Bessie told her sister Elois that Dilworth had been afraid. She had cause to be. In an aerial circus at the same speedway in April of the previous year a woman pilot, Bertha Horchem, was killed during an aerial circus when her plane crumpled in midair and fell 1,200 feet, landing perilously close to a group of spectators. Dilworth might also have heard of the death only the day before of Dallas pilot H. C. Foster in a plane collision near Breckenridge, Texas. In spite of her fear, and undoubtedly with considerable persuasion from Bessie, she made a successful leap off the wing of Bessie's plane at 1,000 feet. Following San Antonio, Bessie did another show at Galveston before moving to Dallas to give a lecture.

Just as Bessie had begun to realize that African American women should be her primary audience, she began to wonder if perhaps the lecture platform might not be more effective than the airfield as a pulpit for preaching her gospel of flight. Consequently, she now lectured more and more frequently, accompanying her talks with clips of the 2,000 feet of film of her performances in Europe and the United States, which she had judiciously preserved and taken with her. Finding that the lectures brought in more income and audiences and were far less expensive than air shows, Bessie spoke at both theaters and schools. But she never charged admission for students, whom she hoped to inspire to become future pilots.

Bessie delivered her first Dallas lecture at the Ella B. Moore Theater, which was managed by Mrs. Moore's husband, Chintz Moore. A local newspaper, describing her as a "very vivacious and loquacious little beauty," also said she was staying in the city with a relative, Lilly V. Beale. None of Bessie's living relatives knew of the Beales. But the city's 1925 city directory lists Mrs. Beale and her husband LeRoy, a janitor, as living at the rear of 208 South Ewing Street.

The hospitality of the Beales in a city with almost no public accommodations for blacks was typical of that shown Bessie in her travels. Wherever she went, fellow African Americans welcomed her, fed her, and housed her. Many of the men were janitors, mailmen, or Pullman porters who, despite their poverty, had found a way to start churches, baseball leagues, and social clubs. Their wives and daughters, often employed as maids and cooks, likewise formed organizations to nurture and protect their communities. Clearly they admired Bessie's accomplishments and supported her aspirations.

Bessie said she would give an aerial exhibition in Dallas on August 27, weather permitting. If she did, there is no evidence of it. If she didn't, it might well have been because she could not borrow a plane. In fact, her main purpose in coming to Dallas had been to buy a plane of her own at nearby Love Field, a veritable shopping mall of used aircraft. Located only seven miles from the present city center, Love Field was then surrounded by cotton fields and ranches. A depot for the storage, dismantling, and sale of surplus aircraft, it was a good place to buy a Jenny for $400 or less. The runway, bordered by a line of wooden hangars, was unpaved and the only housing in view a former Army barracks. To get there from Dallas one took a twenty-five-cent ride in the topless jitney bus that met incoming trains at the Elm and Pacific railway station.

To Bessie, who was accustomed to being regarded as a curiosity at airfields because of her race and sex, Love Field must have been a pleasant surprise. Behncke at Checkerboard Airdrome in Chicago had been kind to her, but she had never before experienced the casual acceptance she found at Love from men who were more interested in piloting and aircraft than in Bessie's gender or color. In fact, one black man, Louis Manning, was already virtually a permanent resident of Love Field. Identified in the city directory as a "yard man," Manning was accepted as a full-fledged member of Love Field's easy-going fraternity. A former delivery man for a drug store, he had been befriended by Byron Good, who serviced and sold planes with his partner H. C. Foster until the latter's death on the day before Bessie's show in San Antonio. When Bessie met Manning, he was Good's right-hand man, acting as office secretary and receptionist, servicing barnstormers' airplanes, controlling crowds at weekend air shows, and directing the parking of cars. And he answered the office telephone with a cheerful "This is Louis Manning, Good and Foster's colored mechanic!"

Like most of the other young men who worked at the field, Manning lived there, sleeping in a hangar or outdoors under the wing of a plane when the weather was good. One of the group, Arthur Spaulding, said,

> [Manning] joined the social life of the field. Everyone met at Ma and Pa Vencill's home, in the old Officers' Quarters on

the field. They sat around the kitchen table, eating Myrtle Vencill's food. Occasionally a stranger would visit the group and gape at Louis in the crowd. When that happened, someone would invariably yell out, "We're all black people here!"

Bessie did not stay long enough to join the Vencill crowd but Spaulding recalled seeing her talking with the pilots in a little café near the edge of the field. "She was a pretty woman," he said, "wearing a fancy uniform. I never did get a chance to talk to her but the boys who did wouldn't have cared if she was colored or not. It was the same as with Louis [Manning]."

Bessie looked over all the planes for sale before she decided on one at the Curtiss Southwestern Airplane and Motor Company. It was probably another JN-4 with an OX-5 engine. However, she could not take it with her. It needed servicing, and the company wanted full payment for it, more money than she had.

Her next stop was at Wharton, eighty miles southwest of Houston, to give two shows—on September 5 and 6. The *Wharton Spectator*, a white-owned newspaper, declared "there was some fancy flying by Bessie Saturday [for] the gathering of a great number of the colored folks at the rodeo grounds." There was also a jazz band from Houston for a dance on Saturday night. However, the report added, "the crowd was disappointed in the failure of Eliza to make her parachute jump." She had been "taken ill," the paper said. But Bessie told Elois that Dilworth had refused to jump. Perhaps that first leap in San Antonio convinced her that a second would be unwise.

The next day Bessie wired Houston for a pilot to come and fly her plane. The patrons had paid to see a jump and she would do it herself. As much an actress as she was a pilot, Bessie was not letting someone else ring down the curtain on her show. On Sunday, after she had finished her stunt flying, Bessie strapped on the jumper's harness and climbed into the front cockpit of her plane while the pilot from Houston took over the controls in the rear.

Straps on the sides of the harness were attached to a parachute inside a canvas laundry bag tied to the outer first strut of the plane. At 3,000 feet Bessie climbed out of the cockpit and walked out on the wing's catwalk to the leading edge, then along the front of the wing to the parachute. When she jumped, her harness pulled the chute out of the bag and she did a free fall until well away from the plane before pulling the release cord. Being Bessie, she landed in the center of the crowd. "The ovation given her was calculated to make her quit flying to do the more spectacular feat of parachute jumping." This was part of a rave review by a white reporter for a white newspaper about a show attended by a mixed audience.

A few days later, the *Houston Post-Dispatch* reported that the people of Wharton had subscribed money for Bessie to buy a plane of her own that would arrive in Dallas on September 13. The source for the story was obviously two of the show's three managers, Hubert Taylor of Wharton and G. R. M. Newman of Houston, who claimed to be in charge of the fundraising campaign. Speaking to the same paper twenty-four hours later, however, Bessie gave a contradictory version. It was she who was buying the plane, with money derived from her air-show earnings and from "additional flights and parachute jumps" she planned to give. Bessie was not letting Newman, Taylor, or Taylor's brother Emmet, all of whom had an interest in the show, take credit for or make any claim on the plane she meant to get from Love Field.

Bessie gave another show in Waxahachie. Elois did not know or say whether her sister went to look at the old house on Palmer Road, a house that her mother still owned and that Bessie had left more than ten years ago. Presumably she did. If so, she would have walked along red dust roads, passing not only the small homes of Wyatt Street blacks but also the Victorian mansions of white Waxahachie, whose front doors were still barred to her. The owners of those mansions undoubtedly had a say in the arrangements for Bessie's show, which was to be before a mixed audience, blacks in one area, whites in another, with separate admission gates for each race. But Bessie drew the line at the two gates. There would be one entrance only, she said, or she would not perform. The organizers finally agreed so long as the audience was segregated after they came in. This time Bessie had to give in. She needed the money from the show and the passenger rides that would follow to pay for her new airplane at Love Field.

Although neither the white nor the black press took note of it, Elois claimed that Bessie also gave a show in the state capital of Austin, after which she was entertained by the governor, Miriam A. "Ma" Ferguson. A colorful figure from Temple, Texas, "Ma" Ferguson succeeded her husband in office after he was impeached. She was an opponent of prohibition and an advocate of tenant farmers. More important to Bessie, she was a Democrat who neither feared nor collaborated with the powerful Ku Klux Klan. And although blacks had almost no voting clout in Texas, one of Ferguson's first acts in office was to pardon a large number of black prisoners. It is entirely possible that the governor asked Bessie to dinner.

Bessie gave other shows in Texas, often in places too small to support a newspaper and thus unrecorded by the press. But she described them to Elois, who provided a good composite picture of her sister's typical performance. Usually it took place at a fairgrounds or racetrack. At the beginning Bessie appeared in her uniform and stood silently while a prayer was said for her safety. Then she walked to her plane and posed for photographers before climbing into the cockpit and taking off for a series of aerobatics. Upon landing, she climbed out of the cockpit and waved her thanks to the cheering crowd that surged forward, pressing against the rope barriers set up to protect her and her plane.

On one occasion a determined small boy slipped under the ropes and through the hands of her assistants. Bessie beckoned to him to approach her. Once he was in her arms, he said, "Lady, didn't your plane stop up there for a little while?" It had, she told him, in a stall. As he walked away, he smiled smugly at the crowd, now the confidant of a celebrity.

Among Bessie's other admirers was a young white man, Paul McCully, who later became National Governor of the OX-5 Pioneers—pilots who flew planes with the OX-5 engine. "I credit this lady with helping launch my flying career," he said. McCully recalled a group of black aviators landing in the river bottom near Fort Worth. With them was "a beautiful black lady in charge of the group, and between selling ride tickets and helping gas the airplanes, she didn't have much time to talk. She drew almost as much crowd as the fliers," McCully recalled.

> She told us of the difficulties a black person had learning to fly. White women were not considered pilot material and black women—no way. She told us she had to leave the country to find an instructor . . . and that she and her friends were trying to make enough money to open a flying school for blacks. I went back to the field early next morning to find out more about the group but they had departed and I never saw them again.
>
> My father was "dead set" against my flying, until he saw these people perform. On the way home papa was puffing and blowing up the long dusty road out of the river bottom. He stopped for air and, wiping his forehead, said, "Well, son, I guess if they can do it you can too." I soloed one month later.

McCully's account, delivered almost seventy years later, is testimony to the vivid impression Bessie left on the people of Texas who saw her, the majority of them African Americans. That summer was the most successful she had ever had. There were no crashes, no canceled engagements, no irate managers or crowds. Her career was no longer under attack from critic J. A. Jackson, who had left the *Billboard* within days of her first appearance in Houston because his column no longer attracted enough paid advertising to justify it. Another good season on the road might well assure her enough capital to open her aviation school.

CHAPTER 9

Nearing the Goal

From Texas Bessie went home to Chicago for three months—long enough to resolve whatever differences she had had with her sometime agent, D. Ireland Thomas. Through the Theater Owners Booking Agency, Thomas now arranged a series of lectures for Bessie in black theaters in Georgia and Florida. There is no recorded evidence of it, but it is possible he may also have had a hand in setting up some of the exhibitions Bessie was booked to give during her speaking tour of the South.

On the night before Christmas Bessie appeared at Elois's door, her bags packed, ready to leave for the first lecture in Savannah, Georgia. "We had a joyful Christmas Eve," Elois recalled. "We made and wrapped Christmas presents and I hemmed a black taffeta dress of hers. We cooked, tasted, and drank coffee and chatted the whole night through. All of a sudden it was broad daylight and time for Bessie to leave."

Bessie arrived in Savannah during the first week of the new year. "The city is thrilled," the black weekly *Savannah Tribune* proclaimed on its front page. She appeared at a local theater where the paper's reviewer was impressed by her "unlimited confidence" in herself.

Whether due to that unlimited confidence or the reporter's ineptness, the story contained three errors: that Bessie went to France in 1919, was licensed by the French government, and was "the only accredited black flyer in the country." Bessie went to France in 1920, got her license from the nongovernmental Fédération Aéronautique Internationale, and, although the only black woman aviator in the world, was not its only accredited black pilot. Hubert Julian had received his flying license from the National Aeronautic Association in the United States in February 1925. The story was correct, though, in saying Bessie's "main anxiety is to see the number of black aviators increased."

Booked for another lecture in Augusta on January 7, Bessie promised she would return to Savannah in time to give an exhibition flight on Sunday, January 10, at Daffin Park Flying Field, where she would "walk the wings of an airplane and do some other stunts." Bessie did not give that show. All available evidence indicates she could not because she had no plane. There was a show that Sunday at Daffin Park Field but it was given by the nationally known Gates Flying Circus. When Bessie originally promised a performance there, she may have assumed she could borrow a plane from them.

From Atlanta Bessie moved on to Florida, appearing at the Liberty Theater in St. Petersburg first, then at theaters in Tampa and West Palm Beach. The popularity of her lectures soon convinced her that her road to future fame might be paved with celluloid. In pursuit of that possibility she wrote to an African American film producer in Arlington, Florida, as follows:

Tampa, Florida.
Feb. 3 1926

Norman Studios
Arlington, Florida

Dear Sirs:—

I was given your address by Mr. Trumbull, owner of the Liberty Theater in St. Petersburg after I appeared there in person and on the screen with 2 reels showing my flights in Europe and America I have my life work that I want put into pictures I know I have been a success in every house I have played in Chicago and other cities. And with only 2 reel as an added attraction I have titled my play Yesterday—Today & Tomorrow.

If you are enerested which I am sure a few remarks from Mr. Trumbull will give you the correct idea what I will mean to you. and as any intelegent person knows that I am the world's first Col Flyer. man or woman. We have one man now a pilot 9 months in Tulsa if you are inerested I will be willing to go father into the matter with you. I am, and know it, the Most Known Colored person (woman alive) other than the Jazz singers)

Write me what you would like to do about this matter.

Bessie Coleman
c/o 1313 Mairon
Street.
Tampa, Fla.

Obviously interested, film company manager R. E. Norman replied almost immediately in a letter dated February 8:

February 8, 1926

Miss Bessie Coleman
Tampa, Fla.

Dear Miss Coleman:—

In your letter of the 3rd., we note that you wish your life work
put into a film entitled, "YESTERDAY, TODAY and TOMOR-
ROW." There is no doubt that with a picture of five or six
reels, properly acted and full of action with you in the leading
role, would be a good drawing card in the colored theatres.

    Mr. D. Ireland Thomas wrote us several weeks ago along
the lines of your letter, but we didn't hear from him further.

    You may go further into the matter and tell us what your
plans are for making such a picture.

    Yours very truly,

Norman Film Manufacturing Company

(signed) R.E. Norman

Bessie's writing ability was limited but her business acumen was
sharp enough to realize that Norman's "tell us what your plans are"
meant "Are you going to pay for it?" Her reply suggested some sort
of coproduction:

W. Palm Beach Fla
Feb. 23, 1926

My Dear Mr. Norman

I just rec' your letter it was *NOT* forwared to me as it should
have been. Yes Mr. Norman I am more than *SURE* my picture
will go big in Colored houses I *know* this, as a Fact as my two
News reels have drawn in house more so than some Colored
Dramas

    You may know what a Real Film of 5 reels would mean
You only have to ask the Mgr. at some of the Theatres in Fla.
Tampa was not advertized "at all" But in St. Petersburg it was
impossible to Show to all. Have a chance to return.

The picture that I want filmed maybe *we* could get together
on it, yesterday Today and tomorrow It would be better if we
jointly put out the photo, as I am not able to produce it
independty Now if you are enerested let me know also give
me a prise on directing 5 reels and est.
Yours
Bessie Coleman
530 First Street.

Apparently Bessie heard nothing more from Norman and all correspondence stopped.

At some time during those first two months in Florida Bessie met
an Orlando couple, the Reverend Hezakiah Keith Hill and his
wife, Viola Tillinghast Hill. Both respected community activists,
the Hills thought Bessie's message of real importance to African
Americans in Orlando. To ensure that it would be heard, Mrs. Hill
invited Bessie to stay with them at the parsonage of her husband's
church, the Mount Zion Missionary Baptist Institutional Church,
west of downtown Orlando in a neighborhood known as the Callahan section.

For Bessie the spacious house on shady, tree-lined Washington
Street became the center of her life, the idyllic home she had never
had. And Viola Hill became friend, sponsor, and surrogate mother,
presiding over a household that radiated generosity and warmth.
Behind the parsonage Viola kept a large vegetable garden, several
cows, a flock of chickens, and fruit trees, all of them providing
food for a table where members of her husband's congregation were
welcomed. There Bessie met many of Orlando's church workers
and school-board members, devout, hardworking people who
shared whatever they had. Within this bastion of faith Bessie redis-
covered hers, and with the help of the Hills, whom she called
"Mother" and "Daddy," she became a "born-again" Christian,
professing her belief in public.

Bessie had always liked children but, with the exception of her nieces and nephews, had known few of them. In Orlando several children lived near the Hills, all of them in delighted awe over the presence of the woman who could fly. Six of them—Aretha Hill Norwood, Jessie Lee Green Griffin, Edyth Lindsey Crooms, Bernice Cox Wheeler, Guyretha Thomas Courtney, and Felix Cosby— can still recall Bessie's affectionate manner with all the children of the neighborhood.

Aretha Hill, the Reverend Hill's niece, who looked after the Hills' cows and chickens, thought Bessie brought excitement and joy to the household and to the whole neighborhood, and especially to the children. "I would always run to take a quick look at that beautiful lady," she said. Aretha was only one of Bessie's many small admirers. At first, Edyth Lindsey, who lived down the block on Washington Street, used to hide under the elevated porch, "to see her come out of the parsonage. She was a very beautiful woman," Edith recalled. Bernice Cox, who lived next door, agreed.

Before long Bessie's shy admirers were no longer hiding under the porch. Jessie Lee Green said Bessie talked to them frequently.

> She was an important lady, famous in air races, but she gave us her time. I lived the second door down but I had a lot of time and every time I saw Bessie Coleman, I was right there. She was a friendly lady, talking to everyone, even us children. She couldn't get rid of us. We admired everything about her, especially her driving airplanes. She promised us she would give us a ride in the airplane. The last time I saw her she had on her uniform. She was smiling. She had beautiful short hair and a beautiful personality.

Among the boys who were awed by Bessie was Felix Cosby, whose father owned a bar and ice-delivery company. Felix and some of his cronies planned to take flying lessons.

> I was so inspired by her talk and her flying that I immediately decided that when I was old enough I'd become a pilot for the

U.S. Army. Of course, growing boys always change their minds if they want to. When World War II began I changed my mind and decided I wanted to be something else. But you know, I will always, always remember seeing and hearing Bessie Coleman.

Not only boys aspired to take lessons. Guyretha Thomas's father suggested that she learn how to fly from Bessie but her mother, she said, "was one of those kind . . . She didn't believe in things like that . . . So she [Bessie] couldn't give me lessons and I cried and I cried. I couldn't get my mother to gee or haw." Bessie's ability to charm the children stretched beyond the Hills' neighborhood. Soon she was a frequent speaker at black schools in Orlando.

For a small fee or passing the hat, depending where she was, she spoke to adults, too, at churches, theaters, and once in a pool hall (with an escort), "so determined was she to have Negro men become air-minded," said Elois.

Viola Hill soon realized that these collections were never going to be enough to fund an aviation school. However, she disapproved strongly of Bessie's raising money by stunt flying and wanted her to confine herself to lecturing and teaching students. To replace the fluctuating income from air-circus flying, Bessie's new friend suggested that she open a beauty shop in Orlando and helped her to do so. Soon after, Bessie wrote to Elois, "I am right on the threshold of opening a school."

Years later, Viola Hill's sister-in-law Audrey Tillinghast, while cleaning out a family dwelling that her husband Sidney had inherited, discarded a "raggedy rattan chair." Her mother-in-law told her, "That was Bessie Coleman's. Haven't you seen her picture on Viola's piano? That lovely lady standing by her airplane? That was her chair."

"That chair was meant for Bessie's beauty shop," Mrs. Tillinghast said, "and I was sorry I didn't know that before I threw it out."

Neither speaking engagements nor part-time beautician's work put a stop to Bessie's drive to fly and her determination to teach others of her race. Orlando resident Edyth Crooms recalled that once, after making a speech at the Mount Zion church, Bessie took many of her audience to a field where she gave the Reverend and Mrs. Hill a ride "in her plane."

There is no evidence that Bessie owned the plane she used in Orlando. It could have been borrowed or rented from the white pilots who lived there. Pilots tended to be a breed apart, far less racist than the general population, and if a woman as charming as Bessie showed her license and a few press notices to a local owner he might well have loaned or rented her his plane. In Savannah she did not give the show she had promised, probably because she couldn't get a plane. In Tampa and St. Petersburg she appeared in theaters but did not fly. She did give an exhibition in West Palm Beach but the plane she used may not necessarily have been hers. The fact that she had her picture taken in Orlando in front of an airplane does not mean that she owned it.

According to Viola Hill, Bessie had been booked for a parachute jump at the annual flower show of the Orlando Chamber of Commerce but threatened to send her plane back to Texas and withdraw from the show when she learned that it would be for whites only and that African Americans would not be allowed to attend. Principles aside, it was a gutsy and even dangerous stand for her to take. Racial segregation in Florida at the time was rigorously enforced, sometimes sanctioned by law but often not. In Daytona Beach, blacks could go nowhere after dark unless they carried a special pass. In Tampa, vigilante-type "night riders," instigated by real-estate operators, engaged in wholesale attempts to frighten black property owners into selling their property and leaving the city.

Bessie's threat to withdraw from the show worked. Orlando's white businessmen backed down, the Chamber of Commerce agreed to allow blacks, and Bessie performed her jump as promised. But her threat to send her plane back to Texas was typical

Bessie bluster. She had no plane yet and was still trying to raise money for the final payment and delivery of the plane on which she had made partial payment the previous summer and which was still sitting in Dallas at Love Field.

Bessie needed a benefactor badly and finally found one in the person of Edwin M. Beeman, son and sole heir of Harry L. Beeman, the millionaire manufacturer of Beeman Chewing Gum. How they met is not known, but they could easily have done so at a three-day flying show that the *Orlando Evening Reporter Star* sponsored in February, or at a new airfield then under construction at a racetrack on East Winter Park Road. Beeman's father had built a mansion across the road from the racetrack, an estate known as Beeman Park. He also built the San Juan Hotel, which Edwin, widely remembered as a short, pleasant-looking, and well-liked man, managed.

It is hardly surprising that the Beeman-Coleman relationship became a matter of much speculation and gossip. Edwin was rich and married and a member of Orlando's white aristocracy. Bessie was poor and black and, while popular within her own community, was hardly welcome in Beeman Park. Beeman's interest in Bessie was probably platonic, that of a rich young man fascinated by aviation and aviators. Whatever his relationship with her, it was he who gave Bessie the money she needed in order to make final payment on her plane in Dallas and have it flown to Jacksonville, the site of her next engagement, scheduled for May 1.

Before Bessie left she promised the Hills that she would return to make her home in Orlando and that she would give up exhibition flying to lecture and teach. She also told them that the time she had spent with them was the happiest she had ever known. For the first time in her life, she said, she was enjoying the comforts of religion. She left Orlando April 27, traveling by train to Jacksonville to await the arrival of her plane from Dallas. With plane and money problems now a thing of the past and with her career in an upswing, Bessie was ready to fly again.

CHAPTER **10**

The Sky Has
a Limit

When Bessie stepped off the train in Jacksonville, she was met by the publicity chairman of the city's Negro Welfare League, John Thomas Betsch. The 21-year-old Betsch was an attractive fellow, an aviation enthusiast who had, in spite of his youth, convinced League members that Bessie Coleman should be the star attraction at the organization's annual Field Day on May 1.

The charming, gregarious Betsch, son of a German brick mason and an African American woman, was a Howard University graduate just starting out in what would be a lifetime as a community activist, businessman, agent for the Atlanta Life Insurance Company, and Thirty-second-Degree Mason. More than a decade after he met Bessie he would have a daughter, Johnnetta Betsch Cole, the present-day president of Spelman College in Atlanta.

Despite the fact that the May 1 show was being put on by a black organization in a strictly segregated city, Betsch wrangled an advance story from a white newspaper. Having recently performed charity shows in Orlando and Palm Beach, the article said, Bessie that coming Saturday would give an exhibition of her flying skills and do a parachute jump from a speeding plane at 2,500 feet. It pointed out that she had 350 career solo flights to her credit and was "one of the two colored persons [in the world] licensed to fly"—the other being Hubert Julian. And, in what could have been a crude attempt to increase attendance, it contained the first published reference in Florida to the California crash in which she was seriously injured.

As her sponsor, young Betsch had also booked a number of speaking engagements for Bessie at churches and theaters. On the night of April 29 she spoke at the Strand Theater after spending the entire day visiting every African American public school in Jacksonville where she found her most fervent, indeed awe-stricken admirers.

"We thought it was just great," said Marian Johnson, then a 15-year-old student at Stanton High School, the only one for African Americans in Jacksonville. "Imagine, a woman flying a plane!" Schoolmate Frances Tyson readily agreed, adding that Bessie was "small and very pretty in her uniform." Hettie Thompson heard Bessie at the Davis Street Elementary School and said she felt "thrilled that a real flier had come to talk to us," pointing out that "in those days we almost never saw planes." And, almost seventy years after the event, Eugenia Mathews, 11 at the time, recalled how impressive Bessie looked when she appeared at the Darnell-Cookman Elementary School "wearing that smart uniform."

Having earlier in Texas switched her focus of attention from audiences of black men to black women, Bessie now, in Orlando and Jacksonville, was concentrating on children as potential aviators of the future.

While she was preaching her doctrine to the people of Jacksonville, 24-year-old white mechanic William D. Wills was taking off from Love Field in Dallas in the Jenny Bessie had finally purchased. One of the earlier models, it acted accordingly, malfunctioning twice during the twenty-one-hour flight and requiring Wills to make forced landings at the Mississippi towns of Meridian and

Farmingdale for a total of five instead of the three originally scheduled landings.

Wills reached Jacksonville on April 28, landing at Paxon Field, a sixty-acre private field at Edgewood Avenue and Lake City Road (now Enterprise and Broadway). Local pilots W. H. Alexander and Laurie Yonge met him at the field and said later that Wills seemed an excellent pilot but they couldn't understand "how he could get that plane to fly all the way from Dallas" since its OX-5 engine, rated at 90 horsepower, was so worn and so poorly maintained that it could really develop only 60 horsepower at most.

Because of the rigid segregation in Jacksonville, Betsch made arrangements for Wills to room at the Church Street home of another white man, Karl Westerfield, who judged him a "fine young fellow." Wills told his host he had worked his way up at Love Field from helper to mechanic to pilot and was currently employed by the Curtiss Southwestern Airplane and Motor Company, a firm managed by his brother-in-law Capt. S. C. Coons, who was also in charge of surplus sales at Love Field.

The day after the young Dallas man arrived with her plane, Bessie walked into a restaurant where Robert Abbott was dining with friends. The *Defender* editor had been visiting his mother, Flora Abbott Sengstacke, in Savannah the previous week and had continued south to Jacksonville to see friends there. Bessie rushed over to his table, kissed him on both cheeks, and told her friends, "This is the man who gave me my chance. I shall never forget him." Whether the two men had met earlier or whether Wills was actually with Bessie at the time, the Chicago publisher "didn't like the looks of the Texan who was to fly with her, and [told Bessie so] in no unmistakable terms," suggesting that she change her plans and find another partner. Bessie ignored Abbott's advice, however, and decided to go up with Wills the next day for a preliminary flight over the racetrack where the exhibition was to be held and to choose a suitable landing site for her jump on Saturday.

At 6:30 Friday morning Bessie called Wills at Westerfield's and told him that she and Betsch, who would be driving them both to Paxon Field, would come by to pick him up in a few minutes. Wills, Betsch, and Bessie arrived at the field at 7:15. As soon as Wills declared the plane ready, Bessie knelt in prayer beside it. Then she asked Wills to take over the controls in the front cockpit while she sat in the rear so she could study the field for a good jump site. She didn't put on her seat belt because to look down on the field required peering over the edge of the cockpit and she was too short to do so if the belt was fastened.

With Wills at the controls, the JN-4 headed for the racetrack and circled for five minutes at 2,000 feet before climbing to 3,500 feet and turning back to Paxon Field. Wills was cruising at 80 miles an hour when the plane suddenly accelerated to 110 miles an hour, then nose-dived. Eyewitness accounts of the altitude to which the plane descended varied but aviators who were watching said it went into a tailspin at 1,000 feet, then flipped upside-down at 500 feet. Bessie fell out, somersaulting end over end until she hit the ground "with a sickening thud, crushing nearly every bone in her body," 100 yards north of Broadway Avenue, a block from Edgewood Street.

Still strapped in his seat, Wills struggled to regain control of the aircraft but failed. Shearing off the top of a pine tree near the edge of the field, the plane crashed on the adjacent farmland owned by Mrs. W. L. Meadows, more than 1,000 feet from where Bessie's body lay. While Mrs. Meadows's son Raymond called the police, she and a neighbor tried to lift the plane off Wills's body but it was too heavy. John Betsch and several police officers reached the scene at just about the same time. Distraught, his hands shaking with shock, Betsch lit a cigarette to calm his nerves. The match ignited gasoline fumes from the plane, which was immediately engulfed in flames. After the trousers of one of the policemen who was trying to pull Wills's body from the plane also caught fire, Betsch was arrested and taken to the Jacksonville jail, where he was held for several hours.

Even before souvenir hunters stripped the barely cooled skeleton of Bessie's plane of all but its charred engine, officials discovered the cause of the accident. A wrench that either had been left in the plane or worked loose from its fitting slid into the control gears and jammed them. Field pilots Alexander and Yonge said Bessie's plane was "an old wreck" to begin with and noted that the gears of a newer plane would have had protective covering. Their view was supported by J. W. Price, himself the owner of four planes, who said, "We've been taking people up for four months and have not had a single accident" and claimed Bessie's accident "wouldn't have happened" had she been using one of his new-model aircraft.

Before noon that Friday several members of the Jacksonville Negro Welfare League took over for the jailed Betsch, arranging for Bessie's terribly mutilated body to be brought from the field to Lawton L. Pratt's funeral home, an African American establishment. In accordance with Jacksonville's rigid segregation practices, Wills's charred body was taken to the Marcus Conant Company's white mortuary pending the arrival of his brother Connor from Tallahassee.

The next day, Saturday, May 1, with souvenir hunters still picking at the skeleton of the plane, Connor Wills boarded the noon train for Terrill, Texas, with his brother's body. Mourning Bessie, the black community of Jacksonville was plunged into gloom. The fairgrounds were deserted, the lemonade and hot dog stands empty, the sawdust-covered midway silent, the newly painted ferris wheel motionless in the burning Florida sun.

Also canceled was the celebration dance that the young women of the Elite Circle and Girls' DeLuxe Club had scheduled to follow the exhibition. Planned especially in Bessie's honor, the dance had been advertised by cards bearing her picture and inviting guests to "An Aerial Frolic" from 8:00 to midnight for an admission fee of seventy-five cents. That night, instead of attending a dance Bessie's admirers came to the funeral home, which was kept open late for people wishing to pay their respects. The line filed past the closed casket from sundown until long after midnight.

On Sunday Bessie's coffin was moved to the Bethel Baptist Institutional Church to lie in state until 3 P.M., when a memorial service was held under the auspices of the Negro Welfare League and the Pride of Maceo, Daughters of Elks, of which Bessie was a member.

More than 5,000 people attended the service, among them hundreds of schoolchildren who had heard Bessie speak the day before she was killed. Only 2,000 mourners were able to crowd into the Greek revival and Romanesque church at Caroline and Hogan streets. The rest had to stand outside the handsome brick and marble building that would have suited even Bessie's grandest aspirations.

The printed programs bore Bessie's picture, and after soloist Daisy Harding reduced the mourners to tears with Bessie's favorite hymn, "I've Done My Work," three ministers—the Rev. John E. Ford, pastor of the church from 1907 to 1943, the Rev. T. H. B. Walker, and the Rev. Scott Bartley—all gave eulogies. At 7:30 that night, a second memorial service was held at St. Phillip Episcopal Church at Cedar and Union streets. Then Bessie's coffin was put aboard a train to Orlando, where her friends the Hills were arranging still another service.

The Hills were heartbroken. They had already received the last letter she wrote before her death, which contained her promise that she would give up flying and confine herself to the lectern. When Bessie's remains reached Orlando, Viola and Hezakiah brought the casket to their church for a Monday service that was one of the largest ever attended at Mount Zion Missionary Baptist.

On that Monday morning the church was banked with flowers from the Mount Zion's Women's Home Missionary Society, the Mount Pleasant Baptist Church, and the leading black citizens of Orlando. The Reverend Hill spoke of Bessie's newly found faith and her kindness to all who knew her. Ministers from every black church in Orlando read resolutions honoring her, and the voices of the church choir and a quartet rang out with "Lead Kindly Light" and "I've Done My Work."

At the close of the service the mourners—including Viola Hill, who would ride with Bessie on her last journey—accompanied the casket to the station to see it off for Chicago. In the words of one report, "As the body of Miss Coleman was being raised into the baggage car, en route to its final resting place, more than five hundred voices, representing the colored population of the city, hummed sweetly, 'My Country 'Tis of Thee.'"

While the African Americans of Orlando praised and prayed for Bessie, William D. Wills was buried in Terrill, Texas, after services attended by his wife and infant daughter; his parents, Mr. and Mrs. David Wills of Dallas; his sister, Mrs. S. C. Coons, and her husband Captain Coons, also of Dallas; and his brothers Connor and Glenn. There is no record of how many others were there.

Bessie's remains arrived in Chicago the morning of Wednesday, May 5. Already several thousand people were crowded into the Forty-third Street Station to get a glimpse of the casket bearing "The Daring Manicure Girl." Also waiting was a military escort from the African American Eighth Infantry Regiment of the Illinois National Guard, which took the coffin to the South Side funeral home of Kersey, Morsell and McGowan. An estimated 10,000 people filed past the coffin that night and all day Thursday to pay their final respects before it was moved Friday morning to the Pilgrim Baptist Church, scarcely a block away at the corner of Thirty-third Street and Indiana Avenue.

At eleven o'clock that morning a trembling Susan, leaning on the arm of her son John and followed by Bessie's sisters Elois, Nilus, and Georgia, was escorted up the aisle and seated in front. After them came six uniformed pallbearers, overseas veterans of the Eighth Regiment, carrying the flag-draped casket. Fifteen hundred mourners filled the church. Another 3,500 people stood outside on the sidewalk and adjacent streets.

Among the twenty-two pallbearers were Congressman Oscar DePriest and attorney Earl B. Dickerson, who had won a major case before the U.S. Supreme Court which upset real-estate covenants barring blacks from residential areas. Ida Wells Barnett, well-known proponent of and passionate speaker for equal rights, was mistress of ceremonies. And Viola Hill delivered a eulogy, speaking proudly of Bessie's refusal to perform any place where African Americans were not admitted.

Pilgrim Baptist pastor Junius C. Austin delivered the funeral oration, saying of Bessie, "This girl was one hundred years ahead of the Race she loved so well, and by whom she was least appreciated." In answer to complaints he had heard that Bessie accepted too much help from white people, he replied, "Heaven help her if white people had not helped her!"

Bessie was buried that day in Lot 580, Section 9, of Chicago's Lincoln Cemetery at Kedzie Avenue and 123rd Street. One year later the Reverend Austin unveiled a memorial stone at the grave, which read:

> In memory of Bessie Coleman, one of the first American
>   women to enter the field of aviation.
>
> Remembered for her courage and accomplishments.
> She fell 5,300 feet while flying in Jacksonville, Florida,
>   April 30, 1926.
> Presented by the Cooperative Business Men's League, Cook
>   County, and Florida friends.

The donors were wrong about the length of her fall but not about her courage and accomplishments.

Mourning her sister, Georgia sobbed, "Oh, Bessie, you tried so hard." Bessie, who had said during her lifetime, "If I can create the minimum of my plans and desires there shall be no regrets," might well have responded with the words of her favorite hymn:

I've done my work. I've sung my song.
I've done some good.
I've done some wrong.
And I shall go
Where I belong.
The Lord has willed it so.

Or perhaps the strong-willed, warm-hearted, uninhibited Bessie would have preferred the ringing declaration of African American poet Mari Evans's "The Rebel:"

When I
die
I'm sure
I will have a
Big funeral.

Curiosity
seekers

coming to see
if I
am really
Dead

or just
trying to make
Trouble.

Epilogue

Ever since Bessie Coleman died more than sixty years ago, a tenuous but persistent undercurrent of speculation has rippled through African American communities as to the real cause of her death. Some suggest sabotage: envious white fliers who couldn't bear the thought of being outdone by a black woman left the wrench loose in her plane on purpose so it might jam the controls. Others hint at outright murder, picturing Wills as either a spurned lover or jealous colleague who intentionally flipped the Jenny so that Bessie would fall but then was not able to get it back under control.

Bessie wasn't the first nor would she be the last pilot to fall out of an airplane. Ten months earlier Carter Leach, a 19-year-old pilot from Waxahachie, was killed at Love Field when his plane overturned and he slipped from the cockpit, falling 500 feet. Like Bessie, Leach had not fastened his safety belt. Soon after that, pilot Al Johnson fell 1,000 feet from his plane while flying over Mineola, New York. Johnson had loosened his belt to peer over the side when the aircraft hit an air pocket and he bounced out.

Nor was an unsecured wrench unique as the cause of a plane crash. As recently as September 1991 a five-inch wrench left inside the front wheel compartment of a U.S. Navy jet caused the $33-million aircraft to crash off the coast of Virginia Beach. It had come loose on takeoff and lodged in the left engine, jamming it and severing the hydraulic lines.

Bessie's sister Elois may not have started the rumors of sabotage but she certainly gave them soil to grow in by writing that Robert Abbott did not trust Wills. And she enriched that soil by adding, "I cannot reconcile myself to . . . Bessie not wearing her parachute [since] safety first has always been one of her strongest mottoes."

The press, too, may have contributed its share by reporting almost unanimously that the wrench "probably" caused Bessie's accident. The wrench did cause it. Whether it was there accidentally or on purpose is the question. There is not a shred of evidence to support insinuations of sabotage or murder.

At the same time, however, it is not hard to see how the tenor of racial relationships that prevailed at the time (and continues to this day) could give birth to and sustain rumors of chicanery and evil. Throughout her career, the white press, with only rare exceptions, either ignored Bessie or treated her with contempt. No more glaring example of this can be found than the way in which Jacksonville's two leading white newspapers covered her death. Both focused total attention on Bessie's flying companion, white mechanic-pilot Wills. The *Florida Times Union* printed Wills's picture, went into detail about his family life and his having recently become a father, and exalted him as the person who was "teaching Bessie how to fly!" Bessie's name appears only twice—in a photo caption and in the lead, and then merely as identification. Thereafter, and only far down in the story, does the paper refer to her three times as "the Coleman woman" and once simply as "the woman." For its part, the *Jacksonville Journal* did concede that Bessie was "said to be the only negro woman aviatrix in the world." But Wills's name came first and for the rest of the story Bessie was never referred to by name, only as "the woman."

Bessie fared no better in the white press outside of Jacksonville. The *Chicago Tribune* buried the crash story on page 10 in a one-paragraph story from the United Press wire service. The paper never even thought to identify her as a resident, let alone a celebrity of Chicago. And back in Florida, the *Orlando Reporter-Star* used the same United Press story and made no mention of Bessie's recent air show and many speaking appearances in that city.

If the white world ignored Bessie, several black newspapers commented when she died on how even her own people had tended to overlook her and failed to appreciate her importance. A *Dallas Express* editorial upon her death remarked that "there is reason to believe that the general public did not completely sense the size of her contribution to the achievements of the race as such." The *Norfolk Journal and Guide* reminded its readers, "Whether they take to it or not, Miss Coleman has taught our women that they can navigate the air and, like all pioneers, she has built her own monument." It was the only monument she would have, other than her tombstone.

In 1927, when the *New York News* sponsored naming an apartment building at 140th Street in Harlem "Coleman Manor," it simultaneously delivered the following scolding to its readers: "It is regretted that colored Americans have not before now had the gratitude and the vision to raise some enduring monument to this black 'Joan of Arc.' "

In 1931 a group of black pilots from the Chicago area, the Challenger Pilots' Association, led by Cornelius R. Coffey, flew their planes over Lincoln Cemetery and dropped flowers on Bessie's grave. The flyover continued as an annual event but eventually lapsed as the originating pilots retired from flying.

In 1934 Lt. William J. Powell, who had already founded Bessie Coleman Aero Clubs on the West Coast, dedicated his book, *Black*

Wings, to her memory. "Because of Bessie Coleman," he wrote, "we have overcome that which was much worse than racial barriers. We have overcome the barriers within ourselves and dared to dream."

In 1977 a group of black women student pilots from the Chicago-Gary, Indiana, area formed the Bessie Coleman Aviators Club, and in 1980 aviation historian and pilot Rufus A. Hunt revived the flyover of her grave. A decade later, on April 28, 1990, pilots from the Chicago American Pilots Association, along with members of the Detroit and New York affiliates of the Negro Airmen International, joined in the annual flyover. And that same year Mayor Richard M. Daley renamed Old Mannheim Road at O'Hare Airport "Bessie Coleman Drive."

On February 26, 1992, during Black History Month, the Chicago City Council passed a resolution presented by Alderman Arenda Troutman and signed by Mayor Daley which requested that the U.S. Postal Service issue a stamp "commemorating Bessie Coleman and her singular accomplishment in becoming the world's first African American pilot and, by definition, an American legend." The resolution noted in part that even sixty-five years after her death "Bessie Coleman continues to inspire untold thousands, even millions of young persons with her sense of adventure, her positive attitude, and her determination to succeed."

Finally, Mayor Daley issued a proclamation on April 27, 1992, urging "all citizens to be cognizant of the accomplishments of the first African American aviatrix of record." Declaring May 2 to be "Bessie Coleman Day in Chicago" and coming sixty-six years after her death, the proclamation represented the first truly official recognition of her many achievements. On that day, aviation historian Hunt again led the annual memorial flight from Gary (Indiana) Municipal Airport to Lincoln Cemetery, where flowers were dropped on her grave.

Notes

Notes appearing in abbreviated form at first reference appear in full in the bibliography.

Chapter 1 • The Reluctant Cotton Picker

3 "SHE HAD THIRTEEN CHILDREN": U.S. Census of 1880 and 1900, provided by Jean Albright Gilley.

"NEITHER SUSAN NOR HER HUSBAND": Marion Coleman, interviews with author, October 24 and 25, 1990 (hereafter Coleman interviews); and Patterson, *Memoirs*.

BESSIE'S PARENTS: Gilley, from U.S. Census.

"ATLANTA WAS A PLACE": Frank X. Tolbert, *Tolbert's Texas Scrapbook*, Vols. 3, 18, 19. Unpublished. Barker Texas History Center, Austin.

4 MEMPHIS LYNCHING: Giddings, *When and Where I Enter*, 17–20.

PARIS LYNCHING: Ibid., 89.

TEXAS VIOLENCE: *Norton's Union Weekly Intelligencer*, January 10, 1892, pp. 1–3.

5 GEORGE COLEMAN'S LAND PURCHASE: Bernice Hamilton, per-

sonal communication with author, Ellis Associates, Surveyors, Waxahachie, Tex., May 13, 1991.

"WAXAHACHIE WAS A TEEMING HUB": Historic Waxahachie Inc. Driving Tour Map. Historic Waxahachie, Inc., Waxahachie, Texas, 1987.

ELLIS COUNTY RECORDS: Beulah Florence, personal communication with author, from the land deeds of Ellis County, Texas, May 13, 1991; Hamilton, communication with author; Historic Waxahachie, Inc., Driving Tour Map.

"ON HIS QUARTER ACRE": Historic Waxahachie Inc. Driving Tour Map; Annie Pruitt, interview with author, May 14, 1991.

6   BESSIE'S EARLY CHILDHOOD: Patterson, *Memoirs;* and Coleman interviews.

"IN 1894": Gilley, from U.S. Census, 1900.

BESSIE'S FAMILY RESPONSIBILITIES: Patterson, *Memoirs.*

FAMILY MEMBERS: Gilley, from U.S. Census, 1900.

7   BESSIE'S SCHOOL: Pruitt interview.

"AS LATE AS 1922": *The Call* (Kansas City, Mo.), August 26, 1922, p. 1.

"STAR PUPIL": Patterson, *Memoirs.*

"IF HE PROTESTED": Meier and Rudwick, *From Plantation to Ghetto,* 196–214.

"115 LYNCHINGS": Allen, *The Big Change.*

8   "SAVAGE AND TREACHEROUS": *A Memorial and Biographical History of Ellis County, Texas* (Chicago: Lewis Publishing, 1892), 61.

"IN OKLAHOMA": Greenberg, *Staking a Claim,* 21–28.

"NEITHER PIONEER NOR SQUAW": Patterson, *Memoirs;* Coleman interviews.

BESSIE'S BROTHERS: Alberta Lipscombe, interview with author, February 25, 1991 (hereafter Lipscombe interview).

"SHE HAD NO KINFOLK": Jones, *Labor of Love, Labor of Sorrow,* 77–78.

WORKING FOR THE JONESES: Coleman interviews; Patterson, *Memoirs.*

9   BESSIE'S HOMELIFE: Patterson, *Memoirs;* Coleman interviews.

10   "DEPENDING ON THE WEATHER": Jones, *Labor of Love*, 17.
     PICKING COTTON: Jones, *Labor of Love*; Patterson, *Memoirs*;
     Coleman interviews; Taulbert, *Once Upon a Time When We
     Were Colored*, 28.

11   "WE LAID OUR COTTON SACKS ASIDE": Patterson, *Memoirs*.
     BESSIE AS FAMILY ENTERTAINER: Ibid.; Coleman interviews.

12   FUNDRAISING: Patterson, *Memoirs*.
     WORKING AS LAUNDRESS: Coleman interviews.
     UNIVERSITY ENROLLMENT: Jilda Stallworth Motley, interviews
     with author, September 16, 1990, and February 18, 1992. The
     school is now Langston University, from which Bessie's great-
     niece, Jilda Stallworth Motley, the granddaughter of Elois, was
     graduated in 1967.
     LANGSTON: Rampersad, *The Life of Langston Hughes*, 1:9; Mot-
     ley interviews.

13   "CONQUERING HERO": Patterson, *Memoirs*.
     "KEEPING HER PLACE": Jones, *Labor of Love*, 125.
     "ELOIS LATER SAID": Patterson, *Memoirs*.
     "THINKING": Ibid.

14   "LEAVE THAT BENIGHTED LAND": Smith, *Chicago*, 388.
     "IN DALLAS": *Dallas Morning News*, May 17, 1912, pp. 10, 17.
     "AMOUNT TO SOMETHING": Coleman interviews.

Chapter 2 • That Wonderful Town

15   CONDITIONS ON THE TRAIN: Taulbert, *Once Upon a Time When
     We Were Colored*, 141.

16   *BIRTH OF A NATION*: Perrett, *America in the Twenties*, 85.
     "SHUNNED BY ALL OTHER GROUPS": Meier and Rudwick, *From
     Plantation to Ghetto*, 237.
     "A VIRTUAL SAVAGE": Smith, *Chicago*, 392.

17   BESSIE MOVES IN WITH WALTER: *Lakeside Directories, 1912–
     1922*; Coleman interviews.
     PULLMAN CONDITIONS: Travis, *An Autobiography of Black Chi-
     cago*, xviii–xix.
     BESSIE'S BROTHERS: Coleman interviews.
     "BLACK BELT" OF CHICAGO: Meier and Rudwick, *From Planta-*

tion to Ghetto, 235; Frazier, *The Negro Family in the United States,* 227–34.

18 BESSIE'S SISTER-IN-LAW: Coleman interviews.
"IN 1915": Giddings, *When and Where I Enter,* 101.
"ONE SUCH CRITIC": Ibid., 115.
HAIR AND SKIN PREPARATIONS: *Chicago Defender,* June 6, 1925; *Guinness Book of Records* (New York: Facts on File, 1977), 496.

19 "THE DEBATE CONTINUED": *Chicago Defender,* July 4, 1925, pt. 2, p. 1.
"SHE WAS GOING TO BE A BEAUTICIAN": Coleman interviews; Vera Stallworth Buntin, interview with author, September 6, 1991 (hereafter Buntin interview).

20 "BEST AND FASTEST MANICURIST": Patterson, *Memoirs.*
"AS A MANICURIST": *Lakeside Directories.*
"THE STROLL": Travis, *An Autobiography of Black Chicago,* 35–37.
NIGHTCLUBS: Ibid., 36; Travis, *An Autobiography of Black Politics,* 42–44; Coleman interviews.

21 PEKIN CAFÉ: *Birmingham* (Ala.) *Reporter,* September 4, 1920, p. 1, and December 4, 1920, p. 1.
"BESSIE'S BROTHER JOHN": Coleman interviews.
ROBERT ABBOTT: Coleman interviews; Patterson, *Memoirs;* Travis, *An Autobiography of Black Chicago,* 25, 39.

22 JESSE BINGA: Travis, *An Autobiography of Black Chicago,* 25, 37, 39, 209, 219; *Norfolk Journal and Guide,* September 3, 1921, p. 8; *Afro-American* (Baltimore), February 14, 1925, p. 13.
OSCAR DEPRIEST: Travis, *An Autobiography of Black Politics,* 42–44, 61.
"IN ADDITION TO BINGA": Penderhughes, *Race and Ethnicity in Chicago Politics,* 20.

23 BESSIE'S MARRIAGE: Coleman interviews; Buntin interview; Marriage License No. 750728, County Clerk, Bureau of Statistics, Cook County, Illinois; *Lakeside Directories.*

24 "I COULDN'T STAND HIM": Coleman interviews.

"IN 1918": Coleman interviews; Dean Stallworth, interview with author, September 12, 1991 (hereafter Stallworth interview); Arthur Freeman, interview with author, September 25, 1991 (hereafter Freeman interview).

BLACKS IN THE SERVICE: Quarles, *The Negro in the Making of America*, 181–87.

25   "TWENTY PERCENT": Travis, *An Autobiography of Black Chicago*, 18.

"STILL OTHERS HAD BEEN HIRED": Quarles, *The Negro in the Making of America*, 185–87.

MAYOR THOMPSON: Travis, *Black Politics*, 55–61, 70–71.

26   CHICAGO RACE RIOT: Quarles, *The Negro in the Making of America*, 390–99; Smith, *Chicago*, 390–91.

"AS JOE AND HIS MOTHER FLED": Travis, *An Autobiography of Black Chicago*, 25.

"THE RIOTS LEFT 38 DEAD": Ibid., 20.

27   "THAT'S IT!": Coleman interviews.

Chapter 3 • Mlle. Bessie Coleman—Pilote Aviateur

29   TRYING TO FIND A TEACHER: Patterson, *Memoirs*.

ROBERT ABBOTT: *Chicago Defender*, September 6, 1924, p. 2.

30   "TAKING ABBOTT'S ADVICE": Patterson, *Memoirs*; Coleman interviews; *Lakeside Directories*.

31   "AS THE FIRST AFRICAN AMERICAN WOMAN PILOT": Dr. Marjorie Stewart Joyner, interviews with author, October 26 and 28, 1990 (hereafter Joyner interviews).

"AN UNNAMED SPANIARD": *Pittsburgh Courier*, May 8, 1926.

"A LOT OF MEN CALLERS": Coleman interviews.

PASSPORT APPLICATION: Patterson, *Memoirs*; State Department Files, Group 59, National Archives, Washington, D.C.

32   VISAS: State Department Files; U.S. Passport No. 109381, dated November 9, 1920 (courtesy Vera Jean and Thomas D. Ramey, Harvey, Illinois).

"BY THE ENGLISH": *Chicago Defender*, October 8, 1921, p. 2. The reference is to Le Crotoy in the Somme. The mistake may have been the error of a *Defender* reporter.

ACCOUNTS OF BESSIE'S STAY: *Chicago Defender*, October 8, 1921, and numerous subsequent interviews with the U.S. press; Marie-Josèphe de Beauregard, Fédération des Pilotes Européennes, letter dated December 10, 1992.

33 THE NIEUPORT: *All the World's Aircraft*, 172–73.
"STRUCTURAL FAILURE": Tallman, *Flying Old Planes*, 56–61; Ethell, "Wings of the Great War," 64.
"EACH TIME SHE TOOK A LESSON": Tallman, *Flying Old Planes*, 56–61; Ethell, "Wings of the Great War," 64.

34 FAI TEST: Coleman File, National Air and Space Museum Archives, Washington, D.C.; Young and Callahan, *Fill the Heavens with Commerce*, 67.
FAI LICENSE: Coleman file, NASM Archives; *L'Aerophile*, Bulletin Officiel de L'Aero-Club de France, p. 287.
"IN FRANCE": Dr. Wilberforce Williams, *Chicago Defender*, December 10, 1921, p. 4.

35 "BESSIE LEFT FRANCE": Passport No. 109381; *Air Service News*, November 1, 1921, p. 11; *Aerial Age Weekly*, October 17, 1921, p. 125; *New York Tribune*, September 26, 1921.
"USING INDIA AS AN EXAMPLE": *Dallas Express*, October 8, 1921, p. 1.
SHUFFLE ALONG: Chicago Defender, August 27, 1921, p. 5; Haney, *Naked at the Feast*, 34–35.
DEFENDER INTERVIEW: *Chicago Defender*, October 8, 1921, p. 2.

37 "THE GOLIATHS": "L'Aeronautique Marchande aux Répertoire Commercial de l'Industrie Aeronautique Francez" (Paris: Gauthier Villars, 1922), 13, 60.

38 "THIS IS MY MOTHER": Coleman interviews.
STUNTS BY WOMEN: Kathleen Brooks-Pazmany, *United States Women in Aviation* (Washington, D.C.: Smithsonian Institution Press, 1991).

39 MORE TRAINING: *Chicago Defender*, October 18, 1921, p. 2.

Chapter 4 • Second Time Around

41 "SHE WAS THE GUEST": *Chicago Defender*, February 25, 1922, p. 3.

"THIS WAS THE HEART OF HARLEM": *Pittsburgh Courier*, October 11, 1924, p. 13.

42 CHARLES GILPIN: *Afro-American* (Baltimore), March 17, 1922, p. 1.
CHICAGO RACE COMMISSION REPORT: *Chicago Daily News*, October 5, 1922, p. 30.

43 GEORGE HARRIS: *Chicago Defender*, February 25, 1922, p. 3.
SPEAKING IN NEW YORK: Ibid.

44 CENSURING HARRIS AND MOORE: *Washington Bee*, August 20, 1921, p. 6; *Afro-American* (Baltimore), July 20, 1923, p. 7.
NIEUPORT PLANE: *Afro-American* (Baltimore), March 10, 1922, p. 10; Patterson, *Memoirs*.
"WHITE AMERICANS": *Cleveland Gazette*, August 4, 1922, p. 1.
"ONE EXPATRIATE SPOKESMAN": *Los Angeles Times*, January 2, 1923, p. 2.

45 HOLLAND VISA: Passport No. 109381.
FOKKER: Hegener, *Fokker*, 54–57; *Chicago Defender*, May 21, 1924, p. 1; *Aerial Age*, June 5, 1922, p. 292.
"OBVIOUSLY DETERMINED": Passport No. 109381.

46 NEWSREEL: *California Eagle*, April 29, 1923, p. 11.
ROBERT THELEN: Correspondence from aviation historian John Underwood to Arthur W. Freeman. Undated.
LETTER FROM CAPTAIN KELLER: *Chicago Defender*, September 9, 1922, p. 3.

47 REPORTERS: Ibid.
NEW YORK TIMES REPORTER: *New York Times*, August 14, 1922, p. 4.
FEATS IN GERMANY: Ibid.; research by Herr Wolfram Müller in Hamburg and East and West Berlin;

48 DOZEN PLANES ORDERED: *New Age*, August 26, 1922, p. 6.
COUNTRIES BESSIE SAID SHE VISITED: Passport No. 109381.
"THE WORLD'S GREATEST WOMAN FLYER": *Chicago Defender*, September 2, 1922, p. 3.
"HEART-THRILLING STUNTS": *Chicago Defender*, September 9, 1922, p. 3.

49   REPORTS ON THE SHOW: *The Call* (Kansas City), September 15, 1922, p. 6; *New York Times*, September 4, 1922, p. 9. HUBERT JULIAN: *Chicago Defender*, September 9, 1922, p. 3; Louis B. Purnell, "The Flight of the Bumble Bee," *Air and Space Magazine* (October/November 1989): 33–34.

50   EDISON MCVEY: *Amsterdam News*, August 1, 1923. *BILLBOARD:* Afro-American (Baltimore), September 15, 1922, p. 9.

51   "IN MEMPHIS": *The Commercial Appeal* (Memphis, Tenn.), October 12, 1972, and October 12, 1922; *Afro-American* (Baltimore), December 15, 1922, p. 3.

Chapter 5 • Pleasing the Crowds, Alienating the Critics

53   "SOCIAL NOTICES": *Chicago Defender*, September 30, 1922, p. 4. PLANE CRASH: *Cleveland Plain Dealer*, September 1, 1922, p. 1.

54   PRE-SHOW PUBLICITY: *Chicago Defender*, October 2, 1922, p. 2. "AS FAR AS GEORGIA WAS CONCERNED": Coleman interviews; and Patterson, *Memoirs*.

55   TWO-COLUMN ADVERTISEMENT: *Chicago Defender*, September 30, 1922, p. 2. "BECAUSE OF THIS YOUNG ENTREPRENEUR": *Chicago Aviation News*, April 27, 1922, p. 9. SPEED DERBY: *New York Times*, September 6, 1922, p. 17.

56   "MY AUNT'S A FLIER": Freeman interview. "BESSIE PERFORMED": *Chicago Defender*, October 21, 1922, p. 2. "AFTER THE SHOW": *Chicago Defender*, September 30, 1922, p. 4.

57   *BILLBOARD* ARTICLE ON FILM: *Afro-American* (Baltimore), December 1, 1922, p. 13. WALKING OUT ON FILM: *Afro-American* (Baltimore), November 10, 1922, p. 3, and December 1, 1922, p. 13. "JACKSON RETURNED BESSIE'S BLAST": Ibid., November 10, 1922, p. 3, and December 1, 1922, p. 13.

58   "IN AN INTERVIEW WITH JACKSON": Ibid. NATIONAL ASSOCIATION OF COLORED FAIRS: *Norfolk Journal and Guide*, August 26, 1922, p. 8.

59 "As early as July": Ibid., July 8, 1922, p. 1, and September 9, 1922, p. 1.

"eccentric": *Afro-American* (Baltimore), December 1, 1922, p. 13.

Most powerful men: *The Call* (Kansas City), July 10, 1923, p. 1; *Norfolk Journal and Guide*, June 27, 1925, p. 9; *Afro-American* (Baltimore), December 15, 1922, p. 3.

60 Chicago Race Commission report: *Chicago Daily News*, October 5, 1922, p. 30.

"took to flying naturally": *Norfolk Journal and Guide*, December 16, 1922, p. 4.

61 "Blivens": *Norfolk Journal and Guide*, January 14, 1922.

62 "the biological function": *Amsterdam News*, February 4, 1925.

Bessie in Baltimore: *Afro-American* (Baltimore), November 17, 1922, p. 2, and November 10, 1922, p. 3.

Chapter 6 • Forced Landing

63 school for aviators: *Afro-American* (Baltimore), November 12, 1922, p. 3.

64 "drop literature from the clouds": *California Eagle*, March 4, 1923, editorial page.

Free air shows: *Los Angeles Evening Herald*, February 17, 1923, Real Estate section, p. 8; *Los Angeles Times*, May 6, 1923, part L, p. 8.

André Peyre: Hatfield, *Los Angeles Aeronautics*, 23, 53, 57.

Peyre and Earhart: Hatfield, *Los Angeles Aeronautics*, 23, 53, 57, 80.

Frank Hodge: *Los Angeles Times*, February 26, 1923.

65 Coast Tire and Rubber Company tour: *Los Angeles Evening Herald*, February 3, 1923, p. 1.

Rockwell Army Intermediate Depot: Edward L. Leiser, letter to author, September 21, 1991.

66 "Miss Coleman is a neat-appearing young woman": *Air Service Newsletter*, February 20, 1922.

"ENTHUSIASTIC, CHARMING GIRL": *California Eagle*, January 27, 1923, p. 3.

INTERNATIONAL FLYING LICENSE: Ibid.

IN LOS ANGELES: *Dallas Express*, February 10, 1923, p. 1; *California Eagle*, February 3, 1923, p. 9.

67 "THE ONLY HINT OF CRITICISM": *Afro-American* (Baltimore), February 2, 1923, p. 3.

68 AIR SHOW: *Afro-American* (Baltimore), February 11, 1923, p. 1.

69 CRASH: Ibid.

OTHER CRASHES: *Los Angeles Times*, pt. 2, January 19, 1923, p. 6; pt. 5, January 21, 1923, p. 14; and January 31, p. 6; *Houston Post-Dispatch*, September 1, 1925, p. 1.

CROWD REACTION: *California Eagle*, February 10, 1923, p. 1.

70 TELEGRAMS: Ibid.

LECTURE: *Afro-American* (Baltimore), February 18, 1923, p. 1.

71 FAMILY REACTION: Coleman interview, August 1, 1992.

SACHS'S LETTER: *California Eagle*, March 4, 1923, p. 4.

PAID NOTICE: *California Eagle*, March 18, 1923, p. 4.

72 "QUEEN BESS OPENS SCHOOL": *The New Age*, May 8, 1923, p. 4.

CONTRACT: *L'Aviation* magazine, Editeur à Saint-Valery-sur-Somme. F. Paillart à Abbeville, 1988, p. 170.

FILM POSSIBILITY: *Afro-American* (Baltimore), May 11, 1923, p. 1; *California Eagle*, March 4, 1923, p. 4.

73 YMCA LECTURES: *California Eagle*, April 29, 1923, p. 11, and May 5, 1923, p. 5.

Chapter 7 • Grounded

75 POST-CRASH AIR SHOW: *Chicago Defender*, June 23, 1923, p. 2, and September 2, 1923, p. 2; *Columbus Evening Dispatch*, September 2, 1923, p. 9, and September 9, 1923, p. 9.

76 KLAN MEETING: *Columbus Evening Dispatch*, September 2, 1923, p. 10, and September 4, 1923, p. 18.

VISIT TO *DEFENDER* OFFICE: *Chicago Defender*, September 8, 1923, p. 2.

77 AIR SHOW: Ibid., September 22, 1923, p. 2.

UNBYLINED STORY: *Afro-American* (Baltimore), May 2, 1924, p. 4.

MANAGERS: Patterson, *Memoirs; Afro-American* (Baltimore), July 7, 1923, p. 6.

78   "A GOOD LONG REST": Patterson, *Memoirs.*

"NO NICHE IN BLACK CHICAGO": *Chicago Defender,* February 9, 1924.

DR. MARJORIE STEWART JOYNER: Joyner interviews.

79   "THE RIGHT SHADE OF BLACK": Travis, *An Autobiography of Black Chicago,* 72, 73.

80   "YOUR AUNT BESSIE IS VERY PRETTY": Coleman interviews and Lipscombe interview.

NIECES AND NEPHEWS: Coleman interviews; Buntin interview; Stallworth interview.

GEORGIA'S DAUGHTER MARION: Coleman interviews. Marion was taught by nuns from Corpus Christi Church at Forty-ninth and Dr. Martin Luther King Jr. Drive.

81   THE RED DRESS: Ibid.

PRINCE OF DAHOMEY: Patterson, *Memoirs;* Coleman interviews; *Dallas Express,* November 29, 1924; *Afro-American* (Baltimore), March 7, 1922, p. 5; *Fort Worth Star-Telegram,* June 9, 1925, p. 8.

"OTHER MEN FRIENDS": Coleman interviews.

JESSE BINGA: *Afro-American* (Baltimore), February 14, 1925, p. 13.

82   ARTICLE ON FOKKER: *Chicago Defender,* May 27, 1924, p. 1.

HUBERT JULIAN: *Norfolk Journal and Guide,* April 19, 1924, p. 1; *Chicago Defender,* July 12, 1924, p. 1.

Chapter 8 • Texas Triumph

85   NEW HOME BASE: *Houston Post-Dispatch,* May 7, 1925, p. 4.

86   FIRST TEXAS FLIGHT: *Houston Informer,* June 27, 1925, p. 1.

JUNETEENTH: *Afro-American* (Baltimore), June 12, 1925, p. 1.

87   ADVANCE STORIES ON AIR SHOW: *Houston Post-Dispatch,* June 14, 1925, p. 9, and June 18, 1925, p. 1.

"SHOOTING AND CUTTING RAMPAGE": *Houston Post-Dispatch,* June 20, 1925, p. 1.

88   MABEL CODY AND GLADYS ROY: Lomax, *Women of the Air*, 35.
"BIGGEST THRILL": *Houston Informer*, June 27, 1925, p. 1, and
July 11, 1925, p. 6.
JIM CROW LAWS: *Afro-American* (Baltimore), August 17, 1923,
p. 1.

89   "THE BALTIMORE WRITER'S EXPERIENCE": *Houston Informer*,
July 11, 1925, p. 6.
SHOWS IN TEXAS: *Houston Post-Dispatch*, July 5, 1925, p. 7;
Chicago Defender, July 25, 1925, p. 2.

90   INDUCEMENT TO SPECTATORS: *Afro-American* (Baltimore), July
18, 1925, p. 4.
"PUT THE RACE ON THE MAP": *Houston Informer*, July 11,
1925, p. 6.
SHOW AT SAN ANTONIO SPEEDWAY: *San Antonio Express*, August 9, 1925, p. 12; *Dallas Express*, September 29, 1925, p.
1; Patterson, *Memoirs*.
OTHER ACCIDENTS: *Pittsburgh* (Texas) *Gazette*, March 7, 1924,
p. 3; *Fort Worth Star Telegram*, August 9, 1925, p. 1.
DILWORTH'S JUMP: *Dallas Express*, August 29, 1925, p. 1.

91   SCHEDULE OF LECTURES: Ibid.
THE BEALES: Ibid.; City Directory of Dallas, Texas (Dallas:
John F. Worley Directory Co., 1925), 1385, 1974.
AERIAL EXHIBITION IN DALLAS: *Dallas Express*, August 28,
1925.

92   LOVE FIELD: Al Harting, "A Love Affair," *Westward Magazine*
(July 15, 1984): 14–21; *Dallas Times-Herald*, July 15, 1984,
pp. 14–21.
LOUIS MANNING: Arthur Spaulding, interviews with author,
May 15 and 16, 1991 (hereafter Spaulding interviews).

93   DILWORTH REFUSES TO JUMP: *Wharton Spectator*, September
11, 1925, p. 6.
BESSIE'S JUMP: Ibid.; Spaulding interviews.

94   MONEY FOR PLANE: *Houston Post-Dispatch*, September 9,
1925, p. 11, and September 10, 1925, p. 5.
SHOW IN WAXAHACHIE: *Waxahachie Daily Light*, April 30,
1926, p. 1; Patterson, *Memoirs*.

95   GOVERNOR FERGUSON: Jordan and Heron, *A Self-Portrait*, 21.
"ON ONE OCCASION": Patterson, *Memoirs*.

96   PAUL MCCULLY: Paul McCully, letter to author, December 28, 1991.
J. A. JACKSON: *Norfolk Journal and Guide*, June 27, 1925, p. 9.

Chapter 9 • Nearing the Goal

97   SERIES OF LECTURES: R. E. Norman, correspondence with Bessie Coleman (February 8, 1926), Manuscript Department, Lilly Library, Indiana University–Bloomington.
VISIT WITH ELOIS: Patterson, *Memoirs*.
"THE CITY IS THRILLED": *Savannah Tribune*, January 7, 1926, p. 1.

98   JULIAN'S LICENSE: *Pittsburgh Courier*, March 14, 1925, p. 9.
"WALK THE WINGS OF AN AIRPLANE": *Savannah Tribune*, January 7, 1926, p. 1.
GATES FLYING CIRCUS: *Savannah Morning News*, January 10, 1926, p. 10.

99   LETTERS: Norman-Coleman correspondence.

101   THE HILLS: Audrey Tillinghast, interviews with author, April 21 and 22, 1991.

102   NEIGHBORHOOD CHILDREN: Tape-recorded interviews provided by Audrey Tillinghast. Subsequent interviews, by the author with Mrs. Tillinghast, were conducted April 21–22, 1991 (hereafter Tillinghast interviews).

103   "SO DETERMINED WAS SHE": Patterson, *Memoirs*.
BESSIE'S CHAIR: Tillinghast interviews.

104   EDYTH CROOMS: Ibid.
THREAT TO WITHDRAW: *Chicago Defender*, May 15, 1926, p. 2.
SEGREGATION IN FLORIDA: *Pittsburgh Courier*, April 10, 1926, p. 2; *Norfolk Journal and Guide*, January 16, 1926, p. 1.

105   EDWIN BEEMAN: *Evening Reporter Star*, February 29, 1926, p. 1.
FINAL PAYMENT ON PLANE: *Chicago Defender*, May 8, 1926, p. 1.
COMFORTS OF RELIGION: Patterson, *Memoirs*.

Chapter 10 • The Sky Has a Limit

107 JOHN THOMAS BETSCH: *Florida Times Union*, May 1, 1926, sect. 2, p. 1; *Chicago Defender*, May 8, 1926, p. 1; Dr. Johnnetta Betsch Cole, letter to author, November 27, 1991. ADVANCE STORY: *Jacksonville Journal*, April 28, 1926, p. 11.

108 STRAND THEATER: *Atlanta Independent*, May 6, 1926, p. 1. SCHOOLCHILDREN'S IMPRESSIONS: Interviews with author: Marian Johnson Jeffers, Hettie Thompson Mills, and Eugenia Mathews Brown, all October 17, 1991; Frances Tyson Johnson, interviews with author, October 17 and 24, 1991. WILLS FLYING BESSIE'S PLANE: *Jacksonville Journal*, April 30, 1926, p. A24.

109 WILLS STAYING WITH WESTERFIELD: *Florida Times Union*, May 1, 1926, sect. 2, p. 1; Spaulding interviews. MEETING ABBOTT: Patterson, *Memoirs*. PRELIMINARY FLIGHT: *Jacksonville Journal*, April 30, 1926, p. A24.

110 FLIGHT AND ACCIDENT: *Florida Times Union*, May 1, 1926, sect. 2, p. 1; *Chicago Defender*, May 8, 1926, p. 1; *Dallas Express*, May 15, 1926, p. 1; unidentified newspaper clipping on microfilm, FSN SC #000386-1, Schomburg Center for Research in Black Culture, New York Public Library; *Amsterdam News*, May 8, 1926, p. 1. BETSCH'S ARREST: *Florida Times Union*, May 1, 1926, sect. 2, p. 1; *Jacksonville Journal*, April 30, 1926, p. A24.

111 CAUSE OF CRASH: *Jacksonville Journal*, April 30, 1926, p. A24, and May 1, 1926, p. 1. WILLS'S BODY: *Florida Times Union*, May 2, 1926, p. 2; *Jacksonville Journal*, May 1, 1926, p. 1. FAIRGROUNDS: *Chicago Defender*, May 8, 1936, p. 1. DANCE CANCELED: Eartha M. M. White Collection, Thomas G. Carpenter Library, University of North Florida, Jacksonville. VIEWING: *Florida Sunday Times Union*, May 2, 1926, p. 2.

112 MEMORIAL SERVICE: White Collection; *Florida Sunday Times-*

Union, May 2, 1926, p. 2; *Chicago Defender*, May 8, 1926, p. 1; *Dallas Express*, May 15, 1926, p. 1; Camilla Thompson, interview and tour with author, April 25, 1991.
ORLANDO SERVICE: *Jacksonville Journal*, May 1, 1926, p. 1; *Pittsburgh Courier*, May 15, 1926, p. 1; *Atlanta Independent*, May 13, 1926.

113 WILLS'S FUNERAL: *Dallas Journal*, May 1, 1926, p. 6.
VIEWING AND SERVICE IN CHICAGO: *Norfolk Journal and Guide*, May 8, 1926, p. 1; *Dallas Express*, May 15, 1926, p. 1; *Chicago Defender*, May 15, 1926, p. 1; Julia Whitfield, *A Brief History of the Pilgrim Baptist Church*; Jacqueline Smith, interview with author, October 31, 1990.

114 MEMORIAL STONE: Patterson, *Memoirs*.
"OH, BESSIE": *Chicago Defender*, May 15, 1926, p. 1.
"I'VE DONE MY WORK": White Collection

115 "THE REBEL": Mari Evans, *I Am a Black Woman*. New York: William Morrow & Co., 1970.

Epilogue

117 OTHER ACCIDENTS: *Houston Post-Dispatch*, June 17, 1925, sect. 2, p. 1, and September 13, 1925, p. 10; *Washington Post*, September 2, 1991, p. B3.
"I CANNOT RECONCILE MYSELF": Patterson, *Memoirs*.

118 FOCUS ON WILLS: *Florida Times Union*, May 1, 1926, sect. 2, p. 1; *Jacksonville Journal*, April 30, 1926, p. A24.
"THERE IS REASON TO BELIEVE": *Dallas Express*, May 15, 1926, editorial page.

119 "WHETHER THEY TAKE TO IT OR NOT": *Norfolk Journal and Guide*, May 15, 1926, p. 14.
"BLACK 'JOAN OF ARC' ": *New York News*, July 30, 1927.
FLYOVER: *Chicago Defender*, March 8, 1990, pp. 1, 34–35.

120 "BECAUSE OF BESSIE COLEMAN": Powell, *Black Wings*.
BESSIE COLEMAN AVIATORS CLUB: "They Take to the Sky," *Ebony* (May 1977): 89–97.
REVIVAL OF FLYOVER: *Chicago Defender*, March 8, 1990, pp. 1, 34, 35.
BESSIE COLEMAN DRIVE: Reports and Communications from

city officers, *Journal-City-Council-Chicago*, February 28, 1990, p. 11794.

"BESSIE COLEMAN CONTINUES TO INSPIRE": Resolution, the City Council of the City of Chicago, February 26, 1992.

BESSIE COLEMAN DAY: Office of the Mayor, City of Chicago, proclamation dated April 27, 1992; announcement, Bessie Coleman Annual Memorial Flight, distributed by Rufus Hunt.

Bibliography

Adoff, Arnold. *The Poetry of Black America: Anthology of the Twentieth Century.* Edited by Arnold Adoff. Introduction by Gwendolyn Brooks. New York: Harper and Row, 1973.

Alford, Sterling G. *Famous First Blacks.* New York: Vantage Press, 1974.

All the World's Aircraft. Founded by the late Fred T. Jane. Edited and compiled by C. G. Gray. London: Sampson Low, Marston and Co., 1918; rpt: New York: Arno Press, 1968.

Allen, Frederick Lewis. *The Big Change.* New York: Harper and Row, 1952.

———. *Only Yesterday.* New York: Harper and Row, 1957.

Anderson, Jervis. *This Was Harlem: A Cultural Portrait, 1900–1950.* New York: Farrar, Straus, and Giroux, 1982.

Bacon, Eve. *Orlando: A Centennial History.* Chuluota, Fla.: Mickler House, 1975.

"Bethel Baptist Institutional Church, 1838–1988." Booklet issued in 1988 on the 150th anniversary of the church in Jacksonville, Florida.

Bilstein, Roger, and Jay Miller. *Aviation in Texas.* Austin: Texas Monthly Press, 1985.

137

Birmingham, Stephen. *Certain People: America's Black Elite*. Boston: Little, Brown, 1977.

"Black, Brave and Flying." *Ebony, Jr.*, February 1979, pp. 16–18.

Bulletin Officiel de L'Aero-Club de France, Brevets de Pilots, *L'Aerophile*, September 1–15, 1921.

Bundles, A'Lelia. *Madame C. J. Walker—Entrepreneur*. New York: Chelsea House, 1991.

Caiden, Martin. *Barnstorming*. New York: Duell, Sloan, and Pearce, 1965.

Carisella, P. J., and James W. Ryan. *The Black Swallow of Death*. Boston: Marlborough House, 1972.

"Chicago Colored Girl Learns to Fly." *Aerial Age Weekly*, October 17, 1921.

Clark, Kenneth B. *Dark Ghetto: Dilemmas of Social Power*. New York: Harper and Row, 1967.

Coles, Robert. *Children of Crisis: A Study of Courage and Fear*. Boston: Little, Brown, 1964.

"Colored Aviatrix Bobs Up Again." *Air Service Newsletter*, February 20, 1923.

Dedmon, Emmett. *Fabulous Chicago*. New York: Random House, 1953.

Ethell, Jeffrey L. "Wings of the Great War." *Air and Space* (October/November 1991): 64.

Evans, Sara M. *Born for Liberty: A History of Women in America*. New York: Free Press, 1989.

Ferenbach, T. R. *Lone Star: A History of Texas and Texans*. New York: Macmillan, 1968.

Frantz, Joe B. *Texas: A History*. New York: W. W. Norton, 1976.

Frazier, E. Franklin. *The Negro Family in the United States*. Rev. and abridged. Chicago: University of Chicago Press, 1948.

Giddings, Paula. *When and Where I Enter: The Impact of Black Women on Race and Sex in America*. New York: William Morrow, 1984.

Greenberg, Jonathan. *Staking a Claim: Jake Simmons, Jr. and the Making of an African-American Oil Dynasty*. New York: Atheneum, 1990.

A Guide to Orlando's Afro American Heritage. A special project of the Central Florida Society of Afro American Heritage, Inc. No date.

Haney, Lynn. *Naked at the Feast: A Biography of Josephine Baker*. New York: Dodd, Mead, 1981.

Hardesty, Von, and Dominic Pisano. *Black Wings: The American Black in Aviation*. Washington: Smithsonian Institution Press, 1987.

Hatfield, D. D. *Los Angeles Aeronautics: 1920–1929*. Los Angeles: Hatfield History of Aeronautics Alumni Library, Northrop Institute of Aeronautics, 1973.

Hegener, Henry. *Fokker: The Man and the Aircraft*. Garden City, N.Y.: Garden City Press, Ltd., 1961.

Hetzel, Gary. "A Flight into the Past." *Orange County Historical Quarterly*, June 1986.

Hughes, Langston, Milton Meltzer, and C. Eric Lincoln. *A Pictorial History of Black America*. 5th rev. ed. New York: Crown, 1983.

Ingle, John P. *Aviation's Earliest Years in Jacksonville (1878–1935)*. Jacksonville, Fla.: Jacksonville Historical Society, 1977.

Johnson, James Weldon, ed. *The Books of American Negro Poetry*. Rev. ed. San Diego: Harcourt Brace Jovanovich, 1959.

Jones, Jacqueline. *Labor of Love, Labor of Sorrow: Black Women, Work and the Family from Slavery to the Present*. New York: Basic Books, 1985.

Jordan, Barbara, and Shelby Hearon. *A Self-Portrait*. Garden City, N.Y.: Doubleday, 1979.

King, Anita. "Family Tree: Brave Bessie, First Black Pilot." *Essence:* pt. 1, May 1976; pt. 2, June 1976.

Lakeside Directories of the City of Chicago. Chicago: Chicago Directory Company, 1912–22.

"License for Air Pilots." *Flying*, June 1921.

Lomax, Judy. *Women of the Air*. New York: Dodd, Mead, 1986.

Mayer, Harold M., and Richard C. Wade. *Chicago: Growth of a Metropolis*. Chicago: University of Chicago Press, 1964.

Meier, August, and Elliott Rudwick. *From Plantation to Ghetto*. 3d ed. New York: Hill and Wang, 1976.

Meltzer, Melton. *The Truth about the Ku Klux Klan*. New York: Franklin Watts, 1982.

The Negro Almanac: A Reference Work on the African American. 5th ed. Edited by Harry A. Ploski and James Williams. Detroit: Gale Research, 1989.

"Negro Aviatrix to Tour Country." *Air Service Newsletter*, November 1, 1921.

O'Neil, Paul, and the editors of Time-Life Books. *Barnstormers and Speed Kings*. Alexandria, Va.: Time-Life Books, 1981.

Patterson, Elois. *Memoirs of the Late Bessie Coleman, Aviatrix*. Privately published by Elois Patterson, 1969.

Penderhughes, Dianne M. *Race and Ethnicity in Chicago Politics*. Urbana: University of Illinois Press, 1987.

Perrett, Geoffrey. *America in the Twenties: A History*. New York: Simon and Schuster, 1982.

Peters, Raymond Eugene and Clinton M. Arnold. *Black Americans in Aviation*. San Diego: New World Aviation Academy, Inc., and Clinton M. Arnold, 1975.

Powell, Lt. William J. *Black Wings*. Los Angeles: Ivan Deach, Jr., 1934.

Quarles, Benjamin. *The Negro in the Making of America*. 2d rev. ed. New York: Collier Books, Macmillan, 1969.

Rampersad, Arnold. *The Life of Langston Hughes*. Vol. 1: *1902–41. I, Too, Sing America*. New York: Oxford University Press, 1986.

Smith, Henry Justin. *Chicago: The History of Its Reputation*. Part 2. New York: Harcourt, Brace, 1929.

Stowe, Harriet Beecher. *Uncle Tom's Cabin*. New York: Viking Press, 1982.

Tallman, Frank. *Flying Old Planes*. Garden City, N.Y.: Doubleday, 1973.

Taulbert, Clifton Lemoure. *Once Upon a Time When We Were Colored*. Tulsa, Okla.: Council Oak Books, 1989.

Terrell, Mary Church. *A Colored Woman in a White World*. Washington, D.C.: National Association of Colored Women's Clubs, Inc., 1968.

"They Take to the Sky." *Ebony*, May 1977, pp. 16–18.

Travis, Dempsey J. *An Autobiography of Black Chicago*. Chicago: Urban Research Institute, 1981.

———. *An Autobiography of Black Politics*. Chicago: Urban Research Press, 1987.

Wallace, Michelle. *Black Macho: The Myth of the Super-Woman*. New York: Dial Press, 1978.

Waters, Enoch. *American Diary: A Personal History of the Black Press.* Chicago: Path Press, 1987.

Woodson, Carter G. *The Negro in Our History.* 9th ed. Washington, D.C. Associated Publishers, 1947.

Young, David, and Neal Callahan. *Fill the Heavens with Commerce: Chicago Aviation, 1825–1926.* Chicago. Chicago Review Press, 1981.

AFTERWORD

Knowledge of aviator Bessie Coleman fills me with sunshine, enthusiasm, daring, courage, sadness, hope, joy, triumph, and a firm grip on reality. Ms. Coleman's zeal for life and dedication to her own vision strengthens for each one of us our own tenuous grip on reality simply because she was—because she existed. To me she is that ephemeral daydream of adventure, strength, audacity, and beauty that we all seek, hope, and somehow know must be present in the world. That daydream truly exists as every day we see people making small strides in overcoming obstacles of gender, birthright, race, ethnicity, economics, illness, poor technology, education, societal condemnation, and fear. In our hearts we know there is someone out there who leaps over conventional wisdom to reality in a single bound! Bessie Coleman's leap nourishes the heart and spirit of each of us who "meets" her.

I did not have the pleasure of meeting up with Ms. Coleman until I was already an astronaut. But when I did, it was as if I had known this best friend all my life. I met her one cold winter afternoon at the DuSable Museum of African American History in Chicago in 1992. Artifacts from her life were in a glass display case. There staring at me from the picture on her aviation license was a young woman who at first glance looked startlingly like me. I read of her travels to France to get instruction in flying airplanes, since no one in her own country, the United States, would teach her because she was a woman and black. I was warmed by the optimism and self-confidence she must have had to determine her own life at a time when most people and especially black women were supposed to take the lot they were given in life and accept the limitations others imposed on them.

I grew up in Chicago and as the first and only black woman astronaut in the world and an African American history aficionado, I was embarrassed and saddened that I did not learn of her until my space flight beckoned on the horizon. In fact, I felt cheated. At the time I was thirty-five, a year older than Bessie when she died. I wished I had known her while I was growing up, but then again I think she was there with me all the time. You see I am convinced that the fall from the airplane that killed her did not kill her spirit. And the calculated arrogance and disdain of the country's major newspapers, magazines, and history books that ignored her did not diminish her existence either. Because what Bessie Coleman affirms is the *life* in each of us.

I could have written about how aviation technology has progressed over the last seventy years and that now space is the new frontier. I could have emphasized the need to acknowledge that every group of people in the world has had scientists, explorers, and adventurers and that to make the best of space exploration all people must be given the chance to participate. I could have stood on my soapbox and said, "Make sure that we do not lose over 75 percent of the talent available to aerospace technology in this country because of ignorance, sexism, and racism." But no words from me could touch the elegance of these demands made manifest by the life of Bessie Coleman.

It is tempting to draw parallels between me and Ms. Coleman. On the surface folks could say we were both young black women born in the South who lived in Chicago and became involved in science and technology fields—aviation and aerospace—when women and especially black women whether through commission or omission were traditionally kept from participation. One could speak of the lack of role models and facing adversity. It is easy to get caught up in the airplane, mechanical, flyboy stuff. But what I hope is common to me and Bessie is the smile of adventure, self-determination, and dogged will to see beauty in the world even as ugly things happen around us and to us.

I point to Bessie Coleman and say without hesitation that here is a woman, a being, who exemplifies and serves as a model to all humanity: the very definition of strength, dignity, courage, integrity, and beauty.

It looks like a good day for flying.

Mae Jemison, M.D.

Index

Abbott, Cecilia, 30

Abbott, Robert Sengstacke, 21, 22, 29–30, 31, 39, 43, 48, 53, 79, 81, 109, 118

Accidents, flying, 53, 69, 90, 117, 118

Acosta, Bertrand B. (pilot), 54

Aerial Age Weekly, 35

African Americans, in early twentieth-century U.S.: in armed forces, 24–25; education of, 7; living conditions for, 7, 8, 17–18, 91; living in France, 30, 34, 44–45; violence against, 4, 7, 26; whites' attitudes toward, 16, 25, 42, 44–45, 64. *See also* Jim Crow laws; Lynchings; Segregation

Afro-American (Baltimore), 44, 67, 77, 78, 88

Air Line Pilots' Association, 55

Air Service News Letter, 35, 65–66

Air shows, 38, 48–50, 51, 53–56, 68–69, 75–76, 77, 86–88, 89, 90, 93–95

Alexander, W. H. (pilot), 109, 111

American Pilots Association, 120

American University (Washington, D.C.), 30

Amsterdam News, 61

Amsterdam, 45

Arlington, Fla., 98

Armstrong, Louis, 21, 78

Army Air Service, 55, 69

Associated Negro Press, 60

Atlanta, Tex., 3, 5, 31, 34

Atlantic Aircraft Corporation, 82

Austin, Tex., 95

Baker, Josephine, 36, 45

Baltimore, 58, 62

Stallworth, Dean (nephew), 24, 80
Stallworth, Elois. *See* Coleman, Elois
Stallworth, Eulah B. (niece), 24, 56
Stallworth, Julius (nephew), 24
Stallworth, Lyle Burnett, 24
Stallworth, Vera (niece), 19, 24, 56, 80
Stroll, the, 20, 21, 22, 63
Stunt fliers, 38, 54, 55, 64, 70, 76, 88, 90, 93

Tammany Hall, 44
Tampa, Fla., 98, 104
Taylor, Emmet, 94
Taylor, Hubert, 94
Temple, Tex., 95
Texas and Pacific railroad, 3
The Call (Kansas City), 49
Theater Owners Booking Agency, 97
Thelen, Robert (pilot), 46
Thomas, D. Ireland, 77, 97
Thomas, Roberta Gwendolyn, 30
Thompson, Hettie, 108
Thompson, William Hale, 25
369th American Expeditionary Force, 48
370th Infantry, 25
Tillinghast, Audrey, 103
Travis, Dempsey, 20
Troutman, Arenda, 120
Tubman, Harriet, 9, 11
Turner, Roscoe (pilot), 47
Tyson, Frances, 108

Udet, Gen. Ernst, 46
Uncle Tom's Cabin, 11

United Beauty Schools Owners and Teachers Association, 79
Universal Negro Improvement Association, 81

von Richtofen, Baron Manfred, 54

Walker, Madame C. J., 19, 78
War Cross (France), 25
Washington Bee, 44
Washington, Booker T., 11
Washington, M. C., 77
Waters, Ethel, 21, 35, 78
Waxahachie, Tex.: air show in, 94–95; life in, 5–8, 10–13
West Palm Beach, Fla., 98, 104
Westerfield, Karl, 109
Wharton Spectator, 93
Wharton, Tex., 93, 94
Wheeler, Bernice Cox, 102
Whipper, Leigh, 57
White, Walter F., 81
White, William, 43, 59, 62, 77
Williams, Dr. Wilberforce, 34
Williams, Leon, 57
Wills, William D., 108, 109, 110, 111, 113, 117, 118
Wing-walkers. *See* Stunt fliers
Witteman Aircraft, 82
World War I, 2, 18, 24–25, 33, 45
Wright, Edward, 23

Yonge, Laurie (pilot), 109, 111
Young, Edward (pilot), 77